A Guide to MATLAB®

for Beginners and Experienced Users

Third Edition

Updated for MATLAB® 8 and Simulink® 8

Brian R. Hunt
Ronald L. Lipsman
Jonathan M. Rosenberg
All of the University of Maryland, College Park

CAMBRIDGE
UNIVERSITY PRESS

CAMBRIDGE
UNIVERSITY PRESS

University Printing House, Cambridge CB2 8BS, United Kingdom

Cambridge University Press is part of the University of Cambridge.

It furthers the University's mission by disseminating knowledge in the pursuit of
education, learning, and research at the highest international levels of excellence.

www.cambridge.org
Information on this title: www.cambridge.org/9781107662223

First and second editions © B. Hunt, R. Lipsman, J. Rosenberg, K. Coombes,
J. Osborn, and G. Stuck 2001, 2006
Third edition © B. Hunt, R. Lipsman, and J. Rosenberg 2014

First published 2001
Second edition 2006
Third edition 2014

Printed in the United Kingdom by T J International Ltd. Padstow Cornwall

A catalogue record for this publication is available from the British Library

Library of Congress Cataloguing in Publication data
Hunt, Brian R., author.
A guide to MATLAB : for beginners and experienced users : updated for MATLAB
8 and Simulink 8 / Brian R. Hunt, Ronald L. Lipsman, Jonathan M. Rosenberg, all of the
University of Maryland, College Park. – Third edition.
pages   cm
Includes bibliographical references and index.
ISBN 978-1-107-66222-3 (paperback)
1. MATLAB.   2. Numerical analysis – Data processing.   I. Lipsman, Ronald L., author.
II. Rosenberg, Jonathan M. (Jonathan Micah), 1951– author.   III. Title.
QA297.H86   2014
510.285′53–dc23
2013046166

ISBN 978-1-107-66222-3 Paperback

Additional resources for this publication at
www-math.umd.edu/schol/a-guide-to-matlab.html

Contents

The symbol ☆ denotes a more advanced chapter or section.

Preface

> MATLAB is a high-level language and interactive environment for numerical computation, visualization, and programming. Using MATLAB, you can analyze data, develop algorithms, and create models and applications. The language, tools, and built-in math functions enable you to explore multiple approaches and reach a solution faster than with spreadsheets or traditional programming languages, such as C/C++ or Java®.
> – The MathWorks, Inc.

That statement encapsulates the view of The MathWorks, Inc., the developer of MATLAB®. MATLAB 8 is an ambitious program. It contains hundreds of commands to do mathematics. You can use it to graph functions, solve equations, perform statistical tests, and much more. It is a high-level programming language that can communicate with its cousins, e.g., Fortran and C. You can produce sound and animate graphics. You can do simulations and modeling (especially if you have access not just to basic MATLAB but also to its accessory Simulink®). You can do symbolic computations (if you have the Symbolic Math Toolbox and its included package, MuPAD®). You can prepare materials for export to the internet. In addition, you can use MATLAB to combine mathematical computations with text and graphics in order to produce a polished, integrated, interactive document.

A program this sophisticated contains many features and options. There are literally hundreds of useful commands at your disposal. The MATLAB help documentation contains thousands of entries. The standard references, whether the MathWorks User's Guide for the product, or any of our competitors, contain a myriad of tables describing an endless stream of commands, options, and features that the user might be expected to learn or access.

MATLAB is more than a fancy calculator; it is an extremely useful and versatile tool. Even if you know only a little about MATLAB, you can use it to accomplish wonderful things. The hard part, however, is figuring out which of the hundreds of commands, scores of help pages, and thousands of items of documentation you need to look at in order to start using it quickly and effectively.

That's where we come in.

Why We Wrote This Book

The goal of this book is to get you started using MATLAB successfully and quickly. We point out the parts of MATLAB you need to know without overwhelming you with details. We help you avoid the rough spots. We give you examples of real uses of MATLAB that you can refer to when you're doing your own work. We provide a handy reference to the most useful features of MATLAB. When you've finished reading this book, you will be able to use MATLAB effectively. You'll also be ready to explore more of MATLAB on your own.

You might not be a MATLAB expert when you finish this book, but you will be prepared to become one – if that's what you want. We figure you're probably more interested in being an expert at your own specialty, whether that's finance or physics, psychology or engineering. You want to use MATLAB the way we do, as a tool. This book is designed to help you become a proficient MATLAB user as quickly as possible, so you can get on with the business at hand.

Who Should Read This Book

This book will be useful to complete novices, occasional users who want to sharpen their skills, intermediate or experienced users who want to learn about the new features of MATLAB 8, or who want to learn how to use Simulink or MuPAD, and even experts who want to find out whether we know anything they don't.

You can read through this guide to learn MATLAB on your own. If your employer (or your professor) has plopped you in front of a computer with MATLAB and told you to learn how to use it, then you'll find the book particularly useful. If you are teaching or taking a course in which you want to use MATLAB as a tool to explore another subject – whether in mathematics, science, engineering, business, or statistics – this book will make a perfect supplement.

As mentioned, we wrote this guide for use with MATLAB 8. If you plan to continue using MATLAB 7, however, you can still profit from this book. Virtually all of the material on MATLAB commands in this book applies to all these versions. The primary features of MATLAB 8 not found in earlier versions have to do with the Desktop interface, the Editor/Debugger, and interaction between MATLAB and MuPAD in the Symbolic Math Toolbox.

How This Book Is Organized

In writing, we used our experience to focus on providing important information as quickly as possible. The book contains a short, focused introduction to MATLAB. It contains practice problems (with complete solutions) so you can test your knowledge. There are several illuminating sample projects that show you how MATLAB can be used in real-world applications and an entire chapter on troubleshooting.

The core of this book consists of about 65 pages: Chapters 1–4, and the beginning of Chapter 5. Read that much and you'll have a good grasp of the fundamentals of MATLAB. Read the rest – the remainder of the Graphics chapter as well as the chapters on Programming, Publishing, MuPAD, Simulink, GUIs, Applications, Troubleshooting, and the Glossary – and you'll know enough to do a great deal with MATLAB.

Here is a detailed summary of the contents of the book.

Chapter 1, *Getting Started*, describes how to start MATLAB on various platforms. It tells you how to enter commands, how to access online help, how to recognize the various MATLAB windows you will encounter, and how to exit the application.

Chapter 2, *MATLAB Basics*, shows you how to do elementary mathematics using MATLAB. This chapter contains the most essential MATLAB commands.

Chapter 3, *Interacting with MATLAB*, contains an introduction to the MATLAB Desktop interface. This chapter will introduce you to the basic window features of the application, to the small program files (M-files) that you will use to make the most effective use of the software, and to a few methods for presenting the results of your MATLAB sessions. After completing this chapter, you'll have a better appreciation of the breadth described in the quote that opens this Preface.

Practice Set A, *Algebra and Arithmetic*, contains some simple problems for practicing your newly acquired MATLAB skills. Solutions are presented at the end of the book.

Chapter 4, *Beyond the Basics*, contains an explanation of some of the finer points that are essential for using MATLAB effectively.

Chapter 5, *MATLAB Graphics*, contains a more detailed look at many of the MATLAB commands for producing graphics.

Chapter 6, *MATLAB Programming*, introduces you to the programming features of MATLAB. This chapter is designed to be useful both to the novice programmer and to the experienced programmer.

Chapter 7, *Publishing and M-Books*, contains an overview of the word-processing and desktop-publishing features available in MATLAB 8.

Practice Set B, *Math, Graphics, and Programming*, gives you another chance to practice what you've just learned. As before, solutions are provided at the end of the book.

Chapter 8, *MuPAD*, describes the symbolic mathematics language MuPAD, which is embedded in the Symbolic Math Toolbox. MuPAD can be used as a stand-alone package for doing a large number of calculations, some of which cannot easily be done with MATLAB directly.

Chapter 9, *Simulink*, describes the MATLAB companion software Simulink, a graphically oriented package for modeling, simulating, and analyzing dynamical systems. Many of the calculations that can be done with MATLAB can be done equally well with Simulink.

Chapter 10, *GUIs*, contains an introduction to the construction and deployment of Graphical User Interfaces, that is *GUIs*, using MATLAB. This chapter is a little more advanced than most of the others.

Chapter 11, *Applications*, contains examples, from many different fields, of solutions of real-world problems using MATLAB, Simulink, and MuPAD.

Practice Set C, *Developing Your MATLAB Skills*, contains problems whose solutions use the methods and techniques you learned in Chapters 8–11.

Chapter 12, *Troubleshooting*, is the place to turn when anything goes wrong. Many common problems can be resolved by reading (and rereading) the advice in this chapter.

Next, we have *Solutions to the Practice Sets*, which contains solutions to all the problems from the three Practice Sets. The *Glossary* contains short descriptions (with examples) of many MATLAB commands and objects. Though it is not a complete reference, the Glossary is a handy guide to the most important features of MATLAB. Finally, there is a comprehensive *Index*.

Conventions Used in This Book

We use distinct fonts to distinguish various entities. When new terms are first introduced, they are typeset in an *italic* font. Output from MATLAB is typeset in a `monospaced typewriter` font; commands that you type for interpretation by MATLAB are indicated by a **boldface** version of that font. These commands and responses are often displayed on separate lines as they would be in a MATLAB session, as in the following example:

```
>> x = sqrt(2*pi + 1)

x =
   2.6987
```

Some MATLAB windows have traditional menu bars near the top; in this book, selectable menu items are typeset in a **boldface** font. Submenu items are separated from menu items by a colon, as in **File:Save**. In MATLAB 8, many windows have a new layout featuring a *Toolstrip*, which has tabs that are organized into sections, each of which contains buttons and menus. Tab names in the Toolstrip are typeset in ALL CAPS; section names in SMALL CAPS; and buttons and menu items therein in **boldface**. We shall sometimes indicate a sequence of selections via colons, as in HOME:FILE:**New**. Labels such as the names of windows and (non-Toolstrip) buttons are quoted, in a "regular" font. File and folder names, as well as web addresses, are printed in a `typewriter` font. Finally, names of keys on your computer keyboard are set in a SMALL CAPS font.

We use seven special symbols throughout the book. Here they are, together with their meanings.

☞   *Paragraphs like this one contain cross-references to other parts of the book, or suggestions of where you can skip ahead to another chapter.*

⇨   **Paragraphs like this one contain important notes. Our favorite is "Save your work frequently." Pay careful attention to these paragraphs.**

✓   Paragraphs like this one contain useful tips or point out features of interest in the surrounding landscape. You might not need to think carefully about them on a first reading, but they may draw your attention to some of the finer points of MATLAB if you go back to them later.

☆   Chapters, sections, or problems starting with this symbol are a little more advanced than the rest of the book, and can be skipped on first reading if you wish.

Paragraphs like this discuss features of MATLAB's Symbolic Math Toolbox, used for *symbolic* (as opposed to *numerical*) calculations. If you are not using the Symbolic Math Toolbox, you can skip these sections.

Paragraphs like this discuss features of Simulink. If you are not using Simulink, you can skip these sections.

Paragraphs like this discuss features of the MuPAD program. If you are not using MuPAD, you can skip these sections.

Incidentally, if you are a student and you have purchased the MATLAB Student Version, then the Symbolic Math Toolbox, MuPAD, and Simulink are automatically included with your software, along with basic MATLAB and a few other toolboxes. We will mention some of these other toolboxes (the Image Processing Toolbox, the Optimization Toolbox, and the Statistics Toolbox) later in this book. With either the Professional Version or the Student Version of MATLAB, additional toolboxes are available if purchased separately.

Additional Online Resources

The reader may find further resources to consult in conjunction with this book – including downloadable M-files used in the creation of some of the Solutions to the Practice Sets and the Applications in Chapter 10 – at the following site:

`www-math.umd.edu/schol/a-guide-to-matlab.html`

About the Authors

We are mathematics professors at the University of Maryland, College Park. We have used MATLAB in our research, in our mathematics courses, for presentations and demonstrations, for production of graphics for books and for the Web, and even to help our kids do their homework. We hope you'll find MATLAB as useful as we do, and that this book will help you learn to use it quickly and effectively.

Appreciation. We are indebted to our former colleagues, Kevin Coombes, John Osborn, and Garrett Stuck, who contributed substantially to earlier editions of this book.

Acknowledgment and Disclaimer. We are pleased to acknowledge support of our research by the National Science Foundation, which contributed over many years to the writing of this book. Our work on the third edition was partially supported by NSF Grants 1136277 and 1206159. Any opinions, findings, and conclusions or recommendations expressed in this material are those of the authors and do not necessarily reflect the views of the National Science Foundation.

Brian R. Hunt
Ronald L. Lipsman
Jonathan M. Rosenberg

College Park, Maryland
August, 2013

Chapter 1

Getting Started

In this chapter, we will introduce you to the tools you need in order to begin using MATLAB effectively. These include the following: some relevant information on computer platforms and software; installation protocols; how to launch MATLAB, enter commands and use online help; a roster of MATLAB's various windows; and, finally, how to exit the program. We know that you are anxious to get started using MATLAB, so we will keep this chapter brief. After you complete it, you can go immediately to Chapter 2 to find concrete and simple instructions for using MATLAB to do mathematics. We describe the MATLAB interface more elaborately in Chapter 3.

Platforms and Versions

The latest version of MATLAB, MATLAB 8 (Release 2012b or later) runs on three operating systems: various versions of Microsoft Windows (both 32-Bit and 64-Bit), Mac OS X, and many distributions of Linux (including Ubuntu, Debian, SUSE, and Red Hat Enterprise). For exact system requirements, you can consult:

```
http://www.mathworks.com/support/sysreq/current_release/
```

For definitiveness, we shall assume that the reader is using a Windows computer. In those very few instances where our instructions must be tailored differently for Linux or Mac users, we shall point it out clearly.

✓   We use the word Windows to refer to all flavors of the Windows operating system. Windows 7 or 8, especially the 64-Bit version, is preferred, though you can also use some Windows XP or Windows Vista installations.

This book is written to be compatible with the current version of MATLAB, namely MATLAB 8 (Release 2012b or later). The vast majority of the MATLAB *commands* we describe apply also to MATLAB 7 (Release 2012a or earlier), but in those earlier versions of the software, many features of the MATLAB interface are different. When important differences between different versions arise, we will point them out. We also note that the differences between the MATLAB Professional Version and the MATLAB Student Version are rather minor, and virtually unnoticeable to a beginner, or even a mid-level user. Again, in the few instances where we describe a MATLAB feature that is available only in the Professional Version, we highlight that fact clearly.

1

Installation

If you intend to run MATLAB, especially the Student Version, on your own computer, it is quite possible that you will have to install it yourself. You can easily accomplish this using the product CDs or via web download, as appropriate. Follow the installation instructions as you would with the installation of any new software. At some point in the installation, you may be asked which *toolboxes* you wish to install. Unless you have severe space limitations, we suggest that you install any included in your license that seem of interest to you or that you think you might use at some point in the future. However, for the purposes of this book, you should be sure to include the *Symbolic Math Toolbox*. We also strongly encourage you to install Simulink, which is described in Chapter 9.

Starting MATLAB

You start MATLAB as you would any other software application. In Windows, you access it via the **Start** menu, under a name like **MATLAB R2013b** or **Student MATLAB**. Alternatively, you may have a desktop icon that enables you to start MATLAB with a simple double-click. In Linux, generally you need only type **matlab** in a terminal window, though you may first have to find the `bin` subdirectory of the MATLAB installation directory and add it to your path. Or you may have an icon on your desktop that achieves the task.

However you start MATLAB, you will briefly see a window that displays the MATLAB logo as well as some product information, and then a *MATLAB Desktop* window will launch. That window will contain a *Toolstrip* and several tabs at the top, with five embedded windows just below the Toolstrip. The largest and most important window is the *Command Window* in the center. We will go into more detail in Chapter 3 on the use and manipulation of three other windows: the *Command History Window*, the *Current Folder Browser*, and the *Workspace Browser*. For now we concentrate on the Command Window in order to get you started running MATLAB commands as quickly as possible. At the top of the Command Window, you may see some general information about MATLAB, perhaps some special instructions for getting started or accessing help, but most important of all, you will see a *command prompt* (**>>** or **EDU>>**). If the Command Window is "active," its title bar will be dark, and the prompt will be followed by a cursor (a blinking vertical line). That is the place where you will enter your MATLAB commands (see Chapter 2). If the Command Window is not active, just click in it anywhere. Figure 1.1 contains an example of a newly launched MATLAB Desktop.

✓   MATLAB 7 had a Desktop with similar windows, but with fewer menu options and a different look. If you need to do a fast calculation and do not want to wait for the Desktop to load, there is an option to run MATLAB without the Desktop. (In Linux or from a Windows Command Prompt, the command is **matlab -nodesktop**.) This gives you a Command Window in a simple terminal window with very few "frills," which often runs faster.

Figure 1.1. A MATLAB Desktop.

Typing in the Command Window

Click in the Command Window to make it active. When a window becomes active, its title bar darkens, and a blinking cursor appears after the prompt. Now you can begin entering commands. Try typing **2+2**, then press the ENTER or RETURN key. Next try **factor(123456789)**, and finally **sin(100)**. Your MATLAB Desktop should look like Figure 1.2.

Online Help

MATLAB has extensive online help. In fact, you should be able to become quite proficient with MATLAB using only this book and the online help.

The easiest way to get help on any aspect of MATLAB or its associated toolboxes is to bring up the *Help Browser* by clicking on one of the two "question mark" icons at the top right of the Desktop or by typing into the search box (also located at the top right of the Toolstrip). You can also type **doc** in the Command Window, followed if you like by the name of the command on which you want documentation. Figure 1.3 depicts the MATLAB Help Browser.

The Help Browser is a powerful and easy-to-use aid in finding the information you need on various features of MATLAB. Like any such tool, the more you use it, the more adept you become at its use.

If you enter a word into the **Search Documentation** box of the Help Browser, or if you enter a word into the search box at the top right of the Desktop Toolstrip, you will see a list of possible matches. On the left of the Help Browser, three regions will appear, enabling you to **Refine by Product**, **Refine by Category**, or **Refine by**

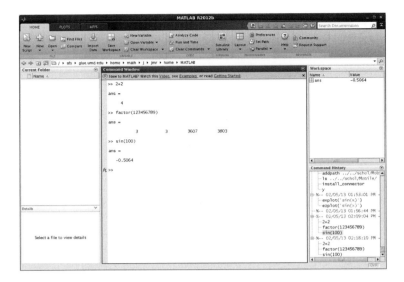

Figure 1.2. The MATLAB Desktop with Several Commands Evaluated.

Type. Refining by product may be important if you have installed a large number of toolboxes, as by default all of them will be searched, but it may be that you are only interested in a single toolbox. For example, a certain command may have a numerical version in MATLAB and a symbolic version in the Symbolic Math Toolbox, and you may want only one of these. Refining by category may be important if the same word is used in several contexts, even within the same toolbox. For example, even within MATLAB, the word "scale" could be used in connection with rescaling graphics, or it could be used in the context of rescaling numbers. Refining by the categories "Graphics" or "Mathematics" would help you find the appropriate help page for your purposes. Refining by type helps you limit your search to MATLAB functions or to examples or to concepts.

✓   In MATLAB Release 2012a or earlier, the Help Browser had a slightly different appearance, with more than one pane. It could still be used in most of the ways we have described, either by searching down the documentation tree or searching for a particular command or key word.

Sometimes the Help Browser produces more detail than you need and it is convenient to have something simpler. If you just want to know the name or syntax of a particular MATLAB command, it is useful to bring up the *Function Browser* by right-clicking in the Command Window and selecting the **Function Browser** menu item.

Finally, for the quickest and simplest version of product help, you can type **help** in the Command Window, followed if you wish by the name of a command or a category of commands. For example, typing **help** at the command prompt will reveal a long list of topics for which help is available. Typing **help general** will give a long list of "general purpose" MATLAB commands. Typing **help factor** will

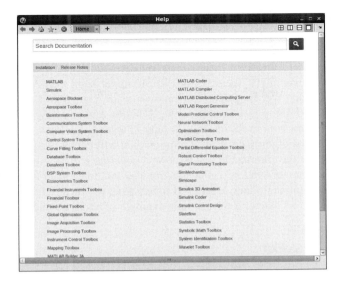

Figure 1.3. The MATLAB Help Browser.

display brief information about the **factor** command. In every instance above, more information than your screen can hold will scroll by. You can use the scroll bar on the right of the Command Window to scroll back up. Alternatively, you can force MATLAB to display information one screenful at a time by typing **more on**. You press the space bar to display the next screenful; type **help more** for details. Typing **more on** affects all subsequent commands, until you type **more off**.

The **lookfor** command searches the first line of every MATLAB help file for a specified string (use **lookfor -all** to search all lines). For example, if you wanted to see a list of all MATLAB commands that contain the word "factor" as part of the command name or brief description, then you would type **lookfor factor**. If the command you are looking for appears in the list, then you can use **help** or the Function Browser on that command to learn more about it.

If you are patient, and not overly anxious to get to Chapter 2, you can type **demo** in the Command Window to bring up a list of videos and scripts that illustrate various aspects of MATLAB and your installed toolboxes.

MATLAB Windows

We have already described the MATLAB Command Window and the Help Browser, and have mentioned in passing the Command History Window, the Current Folder Browser, and the Workspace Browser. These, and several other windows you will encounter as you work with MATLAB, will allow you to do the following: control files and folders that you and MATLAB will need to access; write and edit the small MATLAB programs (M-files) that you will use to run MATLAB effectively; keep track of the variables and functions that you define as you use MATLAB; display and manipulate graphics; and design models to solve problems and simulate processes. Some of these windows launch separately, and some are embedded in the Desktop. You can dock into the Desktop many of those that launch separately by clicking on the downward curved arrow (if there is one) in their tool bars, or by clicking on a button that looks like ⊙ (if there is one) in the upper right corner of the window and selecting **Dock**. You can separate windows inside your MATLAB Desktop out to your computer desktop by clicking on the ⊙ button and selecting **Undock**. These features are described more thoroughly in later chapters. For now, we want to call your attention to the other main type of window you will encounter, namely graphics windows. Many of the commands you issue will generate graphics or pictures. These will appear in a separate window, called a *figure window*. In Chapter 5, we will discuss how to generate and manipulate MATLAB figure windows most effectively.

See Figure 2.1 in Chapter 2 for a simple example of a figure window.

Ending a Session

The simplest way to conclude a MATLAB session is to type `quit` or `exit` (they are indistinguishable) at the prompt. You can also click on the button that generally closes your windows (usually an × in the upper right-hand corner of the Desktop window). *Before* you exit MATLAB, you should be sure to save your work and print any graphics or other files you need. Some strategies for doing so are discussed in Chapter 3.

Mobile Access

Once you've become adept at using MATLAB, you can carry it with you by using the MATLAB Mobile app on a smartphone or tablet device. The app is free to download and use if you have a valid MATLAB license. You can configure it either to connect to your computer, running either the Professional Version or Student Version (R2012a or later) of MATLAB, or to "connect to the cloud" using a MathWorks Account (which you can get with the Professional Version of MATLAB). If you connect to the cloud, then MATLAB Mobile will work even if your computer is turned off. We have tested MATLAB Mobile in both configurations and it does enable you to do many calculations wherever you may be, though you do not have access to the MATLAB Editor or the Simulink graphical environment.

Chapter 2

MATLAB Basics

In this chapter, you will start learning how to use MATLAB to do mathematics. We recommend that you read this chapter while running MATLAB. Try the commands as you go along, and please experiment with variants of the examples we present – if you're not sure whether a command will work, try it!

☞   *For further practice, you can work the problems in* Practice Set A.

Input and Output

The simplest way to input commands to MATLAB is by typing them in the Command Window. MATLAB returns output in two ways: typically, text or numerical output is returned in the same Command Window, but graphical output appears in a separate figure window. Figure 2.1 shows a sample screen, with both a MATLAB Desktop and a figure window. To generate this screen on your computer, first type **1/2 + 1/3**. Then type **ezplot('x^3 - x')**.

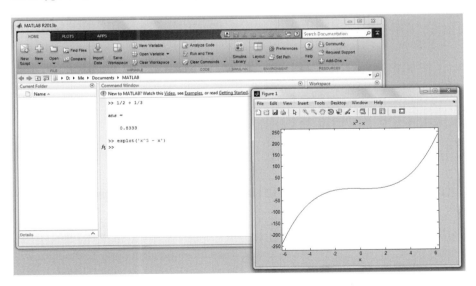

Figure 2.1. MATLAB Windows.

✓   After typing the latter command, you may see MATLAB briefly display the word "Busy" in the lower left corner of its Desktop. When MATLAB is finished running the command, a new command prompt appears.

Arithmetic

As we have just seen, you can use MATLAB to do arithmetic as you would a calculator. You can add with **+**, subtract with **-**, multiply with *****, divide with **/**, and exponentiate with **^**. For example:

```
>> 3^2 - (5 + 4)/2 + 6*3

ans =
   22.5000
```

MATLAB prints the answer and assigns the value to a *variable* called **ans**. If you want to perform further calculations with the answer, you can use the variable **ans** rather than retype the answer. For example, you can compute the sum of the square and the square root of the previous answer as follows:

```
>> ans^2 + sqrt(ans)

ans =
   510.9934
```

Observe that MATLAB assigns a new value to **ans** with each calculation. To do more complex calculations, you can assign each computed value to a variable of your choosing. The assigned value is substituted when you later use the variable. For example:

```
>> u = cos(10)

u =
   -0.8391

>> v = sin(10)

v =
   -0.5440

>> u^2 + v^2

ans =
   1
```

⇨   **Trigonometric functions in MATLAB use radians, not degrees.**

     MATLAB uses double-precision floating-point arithmetic, which is accurate to approximately 15 digits; however, MATLAB displays only five digits by default. To display more digits, type **format long**. Then all subsequent numerical output will have 15 digits displayed. Type **format short** to return to five-digit display.

     MATLAB can also do *exact* arithmetic. For example, it can add the fractions 1/2 and 1/3 symbolically to obtain the correct fraction 5/6. We discuss how to do this in the section *Symbolic Expressions, Variable Precision, and Exact Arithmetic* later in this chapter.

Recovering from Problems

Inevitably, when using any mathematical software system, you are bound to encounter minor glitches. Even while entering simple arithmetic commands, you may accidentally mistype an entry or inadvertently violate a MATLAB rule. In this brief section, we mention three methods for coping with these kinds of problems.

Errors in Input

If you make an error in an input line, MATLAB will normally print an error message. For example, here's what happens when you try to evaluate **3u^2**:

```
>> 3u^2

 3u^2
  |
Error: Unexpected MATLAB expression.
```

The error is a missing multiplication operator **∗**. The correct input would be **3∗u^2**. Note that MATLAB places a marker (a vertical line segment) at the place where it thinks the error might be; however, the actual error may have occurred earlier or later in the expression.

✓ Missing multiplication operators and parentheses are among the most common input errors made by beginning users.

☞ *Recall our discussion of online help in Chapter 1. If you can't decipher an error message caused by a MATLAB command, refer to the online help for that command, or use the search feature of the Help Browser.*

You can edit a previously executed input line as follows. First, redisplay the input line by dragging it with the mouse from the Command History to the Command Window, or by simply typing the UP-ARROW key. Then edit the command and press ENTER or RETURN. You can also use the Command History or the UP- and DOWN-ARROW keys to scroll back and forth through all the commands you've typed in your current and previous MATLAB sessions.

Stopping Execution

If MATLAB gets hung up in a calculation, or seems to be taking too long to perform an operation, you can usually abort it by typing CTRL+C; that is, hold down the key labeled CTRL, or CONTROL, and press C. Make sure the Command Window is highlighted, or else click in the Command Window first. You can also type CTRL+BREAK in Windows or COMMAND+. on a Macintosh. While not foolproof, this is the method of choice when MATLAB is not responding.

Algebraic or Symbolic Computation

Using MATLAB's Symbolic Math Toolbox, you can carry out algebraic or symbolic calculations such as factoring polynomials or solving algebraic equations. Type **help symbolic** to make sure that the Symbolic Math Toolbox is installed on your system.

To perform symbolic computations, you must use **syms** to declare the variables you plan to use to be symbolic variables. Consider the following series of commands:

```
>> syms x y
>> (x - y)*(x - y)^2

ans =
(x - y)^3

>> expand(ans)

ans =
x^3 - 3*x^2*y + 3*x*y^2 - y^3

>> factor(ans)

ans =
(x - y)^3
```

✓   Notice that symbolic output is left-justified, while numerical output is indented. This feature is often useful in distinguishing symbolic output from numerical output.

Although MATLAB makes minor simplifications to the expressions you type, it does not make major changes unless you tell it to. Typing **expand** told MATLAB to multiply out the expression, and **factor** forced MATLAB to restore it to factored form.

MATLAB has a command called **simplify**, which you can sometimes use to express a formula as simply as possible. For example:

```
>> simplify((x^3 - y^3)/(x - y))

ans =
x^2 + x*y + y^2
```

✓   MATLAB has a more elaborate command, called **simple**, that tries more methods than **simplify** does. Try both commands on the trigonometric expression **(sin(x) + cos(x))^2** to compare – you'll have to read the online help for **simple** to completely understand the answer.

Substituting in Symbolic Expressions

When you work with symbolic expressions, you often need to substitute a numerical value, or even another symbolic expression, for one (or more) of the original variables in the expression. This can be done with the command **subs**. For example, if the symbolic expression **w** is defined and it involves the symbolic variable **u**, then the command **subs(w, u, 2)** will substitute the value 2 for the variable **u** in the expression **w**. The result is similar to typing **u = 2** followed by **w**, but using **subs(w, u, 2)** leaves **u** a symbolic variable for future commands. Here are some examples:

```
>> d = 1, syms u v

d =
   1
>> w = u^2 - v^2

w =
u^2 - v^2
>> subs(w, u, 2)

ans =
4 - v^2
>> subs(w, v, d)

ans =
u^2 - 1
>> subs(w, v, u + v)

ans =
u^2 - (u + v)^2
```

✓   When you enter multiple commands on a single line separated by commas, MATLAB evaluates each command and displays the output on separate lines. In the first input line above, the second command produced no output, so we only saw output from the first command.

Symbolic Data, Variable Precision, and Exact Arithmetic

As we have noted, MATLAB uses floating-point arithmetic for its calculations. Using the Symbolic Math Toolbox, you can also do exact arithmetic with symbolic expressions. Consider the following example:

```
>> cos(pi/2)

ans =
   6.1232e-17
```

The answer is written in floating-point format and means 6.1232×10^{-17}. However, we know that $\cos(\pi/2)$ is really equal to 0. The inaccuracy is due to the fact that typing **pi** in MATLAB gives an approximation to π accurate to about 15 digits, not its exact value. To compute an exact answer, instead of an approximate answer, we first create an exact *symbolic* representation of $\pi/2$ by typing **sym('pi/2')**. Then we take the cosine of the symbolic representation of $\pi/2$:

```
>> cos(sym('pi/2'))

ans =
0
```

This is the expected answer.

The quotes around **pi/2** in **sym('pi/2')** create a *string* consisting of the characters **pi/2** and prevent MATLAB from evaluating **pi/2** as a floating-point number. The command **sym** converts the string to a symbolic expression.

✓   The commands **sym** and **syms** are closely related. In fact, **syms x** is equivalent to **x = sym('x')**. Unlike **sym**, **syms** has a lasting effect on its argument. Even if **x** was previously defined, **syms x** clears that definition and renders **x** a symbolic variable – which it remains until (if ever) it is redefined.

As another example, here is how to add $1/2$ and $1/3$ symbolically:

```
>> sym('1/2') + sym('1/3')
```

```
ans =
5/6
```

You can also do *variable-precision arithmetic* with **vpa**. For example, to print 50 digits of $\sqrt{2}$, type:

```
>>  vpa('sqrt(2)', 50)
```

```
ans =
1.4142135623730950488016887242096980785696718753769
```

If you don't specify the number of digits, the default setting is 32. You can change the default with the command **digits**.

⇨   **You should be wary of using sym or vpa on an expression that MATLAB must evaluate before applying variable-precision arithmetic. To illustrate, evaluate the expressions 3^45, vpa(3^45), and vpa('3^45'). The first gives a floating-point approximation to the answer, the second – because MATLAB carries only 16-digit precision in its floating-point evaluation of the exponentiation – gives an answer that is correct only in its first 16 digits, and the third gives the exact answer.**

☞   *See the section* Symbolic and Floating-Point Numbers *in Chapter 4 for details about how MATLAB converts between symbolic and floating-point numbers.*

Vectors and Matrices

MATLAB was written originally to allow mathematicians, scientists, and engineers to handle the tools of linear algebra – that is, vectors and matrices – as effortlessly as possible. In this section, we introduce these tools.

Vectors

A *vector* is an ordered list of numbers. You can enter a vector of any length in MATLAB by typing a list of numbers, separated by commas and/or spaces, inside square brackets. For example:

```
>> Z = [2,4,6,8]

Z =
    2     4     6     8
>> Y = [4 -3 5 -2 8 1]

Y =
    4    -3     5    -2     8     1
```

Suppose that you want to create a vector of values running from 1 to 9. Here's how to do it without typing each number:

```
>> X = 1:9

X =
    1     2     3     4     5     6     7     8     9
```

The notation **1:9** is used to represent a vector of numbers running from 1 to 9 in increments of 1. The increment can be specified as the middle of three arguments:

```
>> X = 0:2:10

X =
    0     2     4     6     8    10
```

Increments can be fractional or negative, for example, **0:0.1:1** or **10:-1:0**.

The elements of the vector **X** can be extracted as **X(1)**, **X(2)**, etc. For example:

```
>> X(3)

ans =
    4
```

To change the vector **X** from a row vector to a column vector, put a prime (**'**) after **X**:

```
>> X'

ans =
    0
    2
    4
    6
    8
   10
```

You can perform mathematical operations on vectors. For example, to square the elements of the vector **X**, type:

```
>> X.^2

ans =
    0     4    16    36    64   100
```

The period in this expression is very important; it says that the numbers in **X** should be squared individually, or *element-by-element*. Typing **X^2** would tell MATLAB to use matrix multiplication to multiply **X** by itself and would produce an error message in this case. (We discuss matrices below and in Chapter 4.) Similarly, you must type **.*** or **./** if you want to multiply or divide vectors element-by-element. For example, to multiply the elements of the vector **X** by the corresponding elements of the vector **Y**, type:

```
>> X.*Y
```

```
ans =
     0      -6      20     -12      64      10
```

Most MATLAB operations are, by default, performed element-by-element. For example, you do not type a period before the addition and subtraction operators, and you can type **exp(X)** to get the exponential of each number in **X** (the matrix exponential function is **expm**). One of the strengths of MATLAB is its ability to efficiently perform operations on vectors.

Matrices

A *matrix* is a rectangular array of numbers. Row and column vectors, which we discussed above, are examples of matrices. Consider the 3×4 matrix

$$A = \begin{pmatrix} 1 & 2 & 3 & 4 \\ 5 & 6 & 7 & 8 \\ 9 & 10 & 11 & 12 \end{pmatrix}.$$

It can be entered in MATLAB with the command:

```
>> A = [1, 2, 3, 4; 5, 6, 7, 8; 9, 10, 11, 12]
```

```
A =
     1      2      3      4
     5      6      7      8
     9     10     11     12
```

Note that the matrix *elements* in any row are separated by commas, and the rows are separated by semicolons. The elements in a row can also be separated by spaces.

If two matrices **A** and **B** are the same size, their (element-by-element) sum is obtained by typing **A + B**. You can also add a scalar (a single number) to a matrix; **A + c** adds **c** to each element in **A**. Likewise, **A - B** represents the difference of **A** and **B**, and **A - c** subtracts the number **c** from each element of **A**. If **A** and **B** are multiplicatively compatible, i.e., if **A** is $n \times m$ and **B** is $m \times \ell$, then their product **A*B** is $n \times \ell$. Recall that the element of **A*B** in the ith row and jth column is the sum of the products of the elements from the ith row of **A** times the elements from the jth column of **B**, i.e.,

$$(\mathbf{A*B})_{ij} = \sum_{k=1}^{m} \mathbf{A}_{ik}\mathbf{B}_{kj}, \ 1 \le i \le n, \ 1 \le j \le \ell.$$

The product of a number **c** and the matrix **A** is given by **c*A**, and **A'** represents the conjugate transpose of **A**. (For more information, see the online help for **transpose** and **ctranspose**.)

A simple illustration is given by the matrix product of the 3×4 matrix **A** above by the 4×1 column vector **Z'**:

```
>> A*Z'
ans =
    60
   140
   220
```

The result is a 3×1 matrix, in other words, a column vector.

☞   *MATLAB has many commands for manipulating matrices. You can read about them in the section* More about Matrices *in Chapter 4 and in the online help.*

Suppressing Output

Typing a semicolon at the end of an input line suppresses printing of the output of the MATLAB command. The semicolon should generally be used when defining large vectors or matrices (such as **X = -1:0.1:2;**). It can also be used in any other situation where the MATLAB output need not be displayed.

Functions

In MATLAB, you will use both built-in functions and functions that you create yourself.

Built-in Functions

MATLAB has many built-in functions. These include **sqrt**, **cos**, **sin**, **tan**, **log**, **exp**, and **atan** (for arctan) as well as more specialized mathematical functions such as **gamma**, **erf**, and **besselj**. MATLAB also has several built-in constants, including **pi** (the number π), **i** (the complex number $i = \sqrt{-1}$), and **Inf** (∞). Some examples:

```
>> log(exp(3))
ans =
   3
```

The function **log** is the natural logarithm, called "ln" in many texts.

```
>> sin(2*pi/3)
ans =
   0.8660
```

To get an exact answer, you need to use a symbolic argument:

```
>> sin(sym('2*pi/3'))
ans =
3^(1/2)/2
```

User-Defined Functions

In this section, we will show how to define your own functions using the operator **@**, which creates what MATLAB calls an "anonymous function." Functions can also be defined in separate files called M-files – see Chapter 3.

Here's how to define the function $f(x) = x^2$ using **@**:

```
>> f = @(x) x^2
```

```
f =
    @(x)x^2
```

Once the function is defined, you can evaluate it:

```
>> f(4)
```

```
ans =
    16
```

As we observed earlier, most MATLAB functions can operate on vectors as well as scalars. To insure that your user-defined function can act on vectors, insert dots before the mathematical operators *****, **/**, and **^**. Thus, to obtain a vectorized version of $f(x) = x^2$, type:

```
>> f = @(x) x.^2
```

Now we can evaluate the function on a vector, for example:

```
>> f(1:5)
```

```
ans =
    1      4      9     16     25
```

You can graph **f** in several ways that we will explore in the section *Graphics* later in this chapter. We conclude this section by remarking that we can also define functions of two or more variables. For example,

```
>> g = @(x, y) x.^2 + y.^2; g(1, 2)
```

results in the answer 5. You can evaluate **g** on vectors of the same size; the command

```
>> g([1 2], [3 4])
```

```
ans =
    10     20
```

gives the values of the function at the points $(1, 3)$ and $(2, 4)$.

✓   Above, we separated two commands on one line with a semicolon instead of a comma. The output of the command preceding the semicolon is suppressed.

Managing Variables

We have now encountered four different types of MATLAB data: floating-point numbers, strings, symbolic expressions, and functions. In a long MATLAB session, it may be hard to remember the names and types of all the variables you have defined. You can type **whos** to see a summary of the names and types, or *classes*, of your currently defined variables. But before you do that, first assign **a = pi, b = 'pi', c**

= **sym('pi')**, and **z = sqrt(-1)**; now type **whos**. Here's the output for the MATLAB session displayed in this chapter:

>> whos

Name	Size	Bytes	Class	Attributes
A	3x4	96	double	
X	1x31	248	double	
Y	1x6	48	double	
Z	1x4	32	double	
a	1x1	8	double	
ans	1x2	16	double	
b	1x2	4	char	
c	1x1	112	sym	
d	1x1	8	double	
f	1x1	32	function_handle	
g	1x1	32	function_handle	
u	1x1	112	sym	
v	1x1	112	sym	
w	1x1	112	sym	
x	1x1	112	sym	
y	1x1	112	sym	
z	1x1	16	double	complex

The variables **A**, **X**, **Y**, **Z**, **a**, **d**, and **z** were assigned numerical data and are reported as class "double". That means that they are arrays of "double-precision" numbers; in this case, the arrays **a**, **d**, and **z** are of size 1×1, i.e., scalars. Also, **ans** is numerical since the last unassigned output was a 1×2 vector. The variable **b** is a string, reported as class "char," while the variables **c**, **u**, **v**, **w**, **x**, and **y** are symbolic. Finally, we see two "function handles," corresponding to the anonymous functions **f** and **g** that we defined. The "Bytes" column shows how much computer memory is allocated to each variable, and the "Attributes" column sometimes indicates a feature of the variable – for example, in the case of **z**, that it is a complex number rather than a real number.

MATLAB commands expect particular classes of data as input, and it is important to know what class of data is expected by a given command; the help text for a command usually indicates the class or classes of input it expects. The wrong class of input usually produces an error message or unexpected output. For example, type **sin(b)** and **2*b** to see what can result from supplying a string when MATLAB expects numerical data.

While **whos** shows information about all defined variables, it does not show the values of the variables. To see the value of a variable, simply type the name of the variable and press ENTER or RETURN. You can also see the values of some variables in the Workspace Browser, which by default is on the right side of the Command Window. Figure 2.2 depicts a Desktop in which the Command Window and the Workspace Browser contain the same information as displayed above. The values of short arrays, strings, and function handles are shown in the Workspace Browser, while for large arrays only the size and class are shown. If you double-click on an array, a new pane appears with the values of the array in a spreadsheet format. This pane is called the Variables Editor, and it allows you both to view and to edit values of a numerical array. You can also open the Variables Editor by clicking on **Open Variable** in the VARIABLE section of the Toolstrip HOME tab.

✓   For Figure 2.2, we minimized the Current Folder Browser and the Command History Window using the pulldown menus obtained by clicking the triangle in the upper right of each pane.

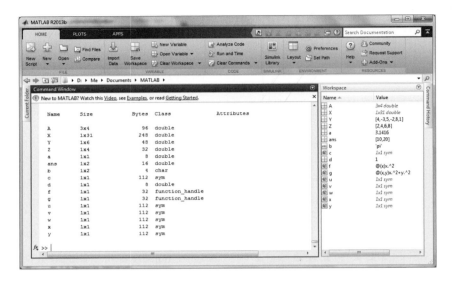

Figure 2.2. Desktop with the Workspace Browser.

Naming and Clearing Variables

A variable name can be any string of letters, digits, and underscores, provided that it begins with a letter and is not ridiculously long. No other punctuation is allowed. MATLAB distinguishes between uppercase and lowercase letters. When using more than a few variables, you should choose distinctive names that are easy for you to remember. For example, you might use **cubicsol** as the name of the solution of a cubic equation.

⇨   **A common source of puzzling errors is the inadvertent reuse of previously defined variables.**

You should also avoid using variable names that coincide with built-in MATLAB functions and commands you might use. For example, if you assign **sec = 60**, then later type **sec(pi)** to find the secant of **pi**, you will get an error message. To fix this problem, you'll need to clear the variable **sec**.

To clear all defined variables, type **clear**, or right-click in the Workspace Browser and select **Clear Workspace**. You can also clear specific variables, either by deleting them in the Workspace Browser or, for example, by typing **clear x y** to clear **x** and **y**. You should generally clear variables before starting a new project. Otherwise, values from a previous calculation can creep into the new calculation by accident.

Solving Equations

You can solve equations involving variables symbolically with **solve**, which is part of the Symbolic Math Toolbox, or numerically with **fzero**. For example, to find the solutions of the quadratic equation $x^2 - 2x - 4 = 0$, type

```
>> syms x; solve(x^2 - 2*x - 4 == 0)

ans =
 5^(1/2) + 1
 1 - 5^(1/2)
```

The answer consists of the exact (symbolic) solutions $1 \pm \sqrt{5}$. To get numerical solutions, type **double(ans)**, or **vpa(ans)** to display more digits.

⇨ **Symbolic equations require a double equals sign, as in the example above, to distinguish them from the usual MATLAB interpretation of a single equals sign as assigning a value to a variable.**

✓ If the right-hand side of the equation to be solved is zero, you can instead input the left-hand side as an expression, and **solve** will set it equal to zero. So above, we could have typed instead **solve(x^2 - 2*x - 4)**.

The input to **solve** can also be a string, and in this case a single equal sign is allowed. Thus, to solve $x^2 = 3x - 7$ you can type:

```
>> solve('x^2 = 3*x - 7')

ans =
 (19^(1/2)*i)/2 + 3/2
 3/2 - (19^(1/2)*i/2)
```

This time, the answer consists of complex numbers $(3 \pm \sqrt{19}i)/2$, where the letter **i** in the answer stands for the imaginary unit $\sqrt{-1}$. Again, to get numerical solutions, use **double** or **vpa**.

The command **solve** can solve higher-degree polynomial equations, as well as many other types of equations. It can also solve equations involving more than one variable. If there are fewer equations than variables, you should specify which variable(s) to solve for. For example, the following command solves $2x - \log y = 1$ for y in terms of x:

```
>> syms y; solve(2*x - log(y) == 1, y)

ans =
exp(2*x - 1)
```

You can specify more than one equation as well. For example:

```
>> [xsol, ysol] = solve(x^2 - y == 2, y - 2*x == 5)

xsol =
 2*2^(1/2) + 1
 1 - 2*2^(1/2)
```

```
ysol =
 4*2^(1/2) + 7
 7 - 4*2^(1/2)
```

There are two ordered pairs (x, y) that solve this system of equations. MATLAB puts their x-values in one vector and their y-values in another vector. The first solution consists of the first value of x together with the first value of y. You can extract these values by typing **xsol(1)** and **ysol(1)**:

```
>> [xsol(1) ysol(1)]
```

```
ans =
[ 2*2^(1/2) + 1, 4*2^(1/2) + 7]
```

The second solution can be extracted with **xsol(2)** and **ysol(2)**.

⇨   **We were careful not to name the outputs x and y in the example above. This would have superseded our earlier declarations syms x and syms y, and caused the solve commands below to produce error messages.**

Note that, in the preceding **solve** command, we assigned the output to the *vector* **[xsol, ysol]**. If you use **solve** on a system of equations without assigning the output to a vector, then MATLAB does not automatically display the values of the solution:

```
>> sol = solve(x^2 - y == 2, y - 2*x == 5)
```

```
sol =
 x: [2x1 sym]
 y: [2x1 sym]
```

To see the vectors of x- and y-values of the solution, type **sol.x** and **sol.y**. To see the individual values, type **sol.x(1)**, **sol.y(1)**, etc.

☞   *In this example, the output of* **solve** *is a* structure. *See the section* Cell and Structure Arrays *in Chapter 6 for more about this data class.*

Some equations cannot be solved symbolically, and in these cases **solve** tries to find a numerical answer. For example:

```
>> solve(sin(x) == 2 - x)
```

```
ans =
1.1060601577062719106167372970301
```

Sometimes there is more than one solution, and you might not get what you expected. For example:

```
>> digits(10), solve(log10(x) == cos(x))
```

```
ans =
- 226.7493799 + 1.678107548*i
```

(Here **log10** is the base-10 logarithm, and we use ten significant digits to shorten the output.) Though the complex number MATLAB reports is a valid solution of the equation, there are also real-number solutions. In fact, the graphs of $\log_{10}(x)$ and $\cos(x)$ are shown in Figure 2.3; each intersection of the two curves represents a solution of the equation $\log_{10}(x) = \cos(x)$.

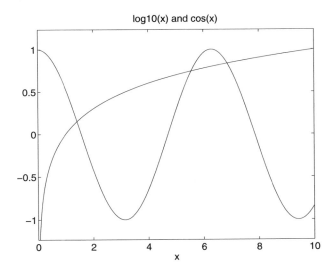

Figure 2.3. Two Intersecting Curves.

You can tell **solve** to consider only real values of x as follows:

```
>> solve(log10(x) == cos(x), 'Real', true)

ans =
5.552112002
```

You can also tell all future symbolic commands to assume **x** is real by typing **syms x real** or **assume(x, 'real')**. In either case, the assumption persists until you clear it with **syms x clear**. On the other hand, the **solve** command above makes the assumption internally and does not alter the future meaning of **x**.

The output above is still only one of the three real solutions shown in Figure 2.3. You can numerically approximate the other solutions using **fzero**, which looks for a zero of a given function near a specified value of x. A solution of the equation $\log_{10}(x) = \cos(x)$ is a zero of the function $\log_{10}(x) - \cos(x)$, so to find an approximate solution near $x = 2$, type:

```
>> fzero(@(x) log10(x) - cos(x), 2)

ans =
   1.4184
```

Replace **2** with **7** to find the largest solution. You can also get these solutions by preceding the **solve** command above with, for example, **assume(0 < x < 2)** or **assume(x > 6)**. See the online help for more possible uses of **assume**.

Graphics

In this section, we introduce MATLAB's two basic plotting commands.

Graphing with `ezplot`

The simplest way to graph a function of one variable is with **ezplot**, which expects a string, a symbolic expression, or a function handle, representing the function to be plotted. For example, to graph $x^2 + x + 1$ on the interval -2 to 2 (using the string form of **ezplot**), type:

```
>> ezplot('x^2 + x + 3', [-2 2])
```

The plot will appear on the screen in a new window labeled "Figure 1".

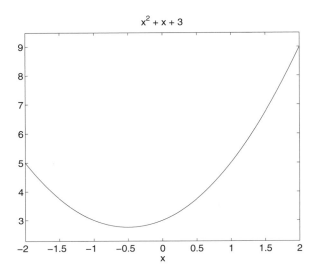

Figure 2.4. The Parabola $y = x^2 + x + 3$ on the Interval $[-2, 2]$.

Using a symbolic expression, you can produce the plot in Figure 2.4 with the following input:

```
>> syms x, ezplot(x^2 + x + 3, [-2 2])
```

However, in this case the title MATLAB puts above the graph may be different than you intended, because MATLAB often rearranges symbolic expressions. Finally, you can use an anonymous function as the argument to **ezplot**, as in:

```
>> ezplot(@(x) x.^2 + x + 3, [-2 2])
```

✓   Graphs can be misleading if you do not pay attention to the axes. For example, the input **ezplot('x^2 + x + 5', [-2 2])** produces a graph that looks identical to the previous one, except that the vertical axis has different tick marks (and MATLAB assigns the graph a different title).

Modifying Graphs

You can modify a graph in a number of ways. You can change the title above the graph in Figure 2.4 by typing (in the Command Window, not the figure window):

```
>> title 'A Parabola'
```

The same change can be made directly in the figure window by making the selection **Axes Properties...** from the **Edit** menu at the top of the figure window. (Just type the new title in the box marked "Title.") You can add a label on the vertical axis with **ylabel** or change the label on the horizontal axis with **xlabel**. Also, you can change the horizontal and vertical ranges of the graph with the **axis** command. For example, to expand the vertical range to the interval from 0 to 10, type:

```
>> axis([-2 2 0 10])
```

The first two numbers are the range of the horizontal axis; both ranges must be included, even if only one is changed. If you unintentionally cut off part of a graph, the missing part is not forgotten by MATLAB. You can readjust the ranges, or type **axis tight** to automatically set the ranges to include the entire graph.

To make the shape of the graph square, type **axis square**; this also makes the scale the same on both axes if the x and y ranges have equal length. For ranges of any lengths, you can force the same scale on both axes without changing the shape by typing **axis equal**. Generally, this command will expand one of the ranges as needed. See the online help for more possibilities.

You can make many more changes directly in the figure window, as you will see if you explore the pull-down menus on the figure window's tool bar. The advantage of modifying figures with MATLAB commands is that the results are easily reproducible, for example by saving your commands in an M-file (see Chapter 3) in order to make a similar graph later on. To close the figure, type **close**, or simply click on the × in the upper right-hand corner of the window.

☞   *See Chapter 5 for more ways to manipulate graphs.*

Graphing with **plot**

The **plot** command works on vectors of numerical data. The syntax is **plot(X, Y)**, where **X** and **Y** are vectors of the same length. For example:

```
>> X = [1 2 3]; Y = [4 6 5]; plot(X, Y)
```

The result is shown in Figure 2.5. The **plot** command above considers the vectors **X** and **Y** to be lists of the x and y coordinates of successive points on a graph, and joins the points with line segments. So, MATLAB connects $(1, 4)$ to $(2, 6)$ to $(3, 5)$.

To plot $x^2 + x + 3$ on the interval from -2 to 2, we first make a vector **X** of x-values, and then type **plot(X, X.^2 + X + 3)**. (The **.** here is essential, since **X.^2** represents the element-by-element square of **X**, not the square by matrix multiplication.) We need to use enough x-values to ensure that the resulting graph drawn by "connecting the dots" looks smooth. We'll use an increment of 0.01. Thus a recipe for graphing the parabola is:

```
>> X = -2:0.01:2; plot(X, X.^2 + X + 3)
```

The result is similar to Figure 2.4, but with no title nor axis label. Note that we used a semicolon to suppress printing of the 401-element vector **X**.

☞   *We describe more of MATLAB's graphics commands in Chapter 5.*

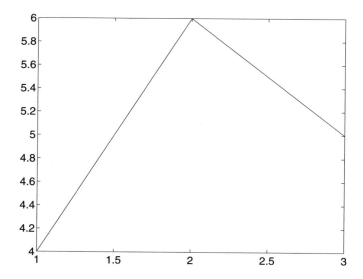

Figure 2.5. Plotting Line Segments.

Plotting Multiple Curves

Each time you execute a plotting command, MATLAB erases the old plot and draws a new one. If you want to overlay two or more plots, type **hold on**. This command instructs MATLAB to retain the old graphics and draw any new graphics on top of the old. It remains in effect until you type **hold off**. Here's an example using **ezplot**:

```
>> ezplot(@log10, [0 10])
>> hold on
>> ezplot(@cos, [0 10])
>> hold off
>> title 'log10(x) and cos(x)'
```

The result is shown in Figure 2.3 earlier in this chapter. Here we used **@** to turn existing MATLAB functions into function handles. The commands **hold on** and **hold off** work with all graphics commands.

With **plot**, you can plot multiple curves directly. For example:

```
>> X = 0:0.01:10; plot(X, log10(X), X, cos(X))
```

Note that the vector of x-coordinates must be specified once for each function being plotted.

✓   Using a single **plot** command has the advantage that the curves are automatically drawn with different colors.

☞   *See Chapter 5 for more on controlling the color and style (e.g., dashed) of curves.*

Chapter 3

Interacting with MATLAB

In this chapter, we describe an effective procedure for working with MATLAB, and for preparing and presenting the results of a MATLAB session. In particular, we discuss some features of the MATLAB interface and the use of *M-files*. We describe how to write a simple loop as a precursor to a more robust discussion of MATLAB programming (in Chapter 6) that will enable you to make the most efficient use of M-files. We introduce the MATLAB publishing feature, which produces formatted output. It is the most effective method for presenting MATLAB results, although we also describe several alternate methods. Finally, we give some simple hints for debugging your M-files.

The MATLAB Interface

MATLAB's basic interactive interface, which launches when you start the program, is called the MATLAB *Desktop*. Embedded inside it is the Command Window that we described in Chapter 2.

The Desktop

By default, the MATLAB Desktop (Figure 1.1 in Chapter 1) contains four windows inside it, the Command Window in the center, the Current Folder Browser on the left, the Workspace Browser in the upper right, and the Command History Window in the lower right. Notice that if you minimize any of the latter three windows, then tabs appear that allow you to alternate between them. You can change which windows are currently visible by clicking on the **Layout** button in the ENVIRONMENT section of the HOME tab in the *Toolstrip* at the top of the Desktop (more on the Toolstrip below), and you can adjust the sizes of the windows by dragging their edges with the mouse. You can click in any one of the four windows to work in it; if you do, its title bar becomes dark (perhaps blue) indicating that it is the active window. The Command Window is where you type the commands and instructions that cause MATLAB to evaluate, compute, draw, and perform all the other wonderful magic that we describe in this book. We will discuss the other windows in separate sections below.

The MATLAB Desktop includes a *Toolstrip*, which is a combination and augmentation of the Menu Bar and Tool Bar that appeared in older versions of the software. The first tab on the Toolstrip, labelled HOME, contains icons that give quick access to the basic features you will use to manipulate files, manage variables, run code, set preferences, and access help. These icons are grouped into "Sections" according to their function. There are two more tabs on the Toolstrip, entitled PLOTS and APPS.

25

The icons in the PLOTS tab provide immediate and convenient access to the most common graphical commands that you will use to generate MATLAB plots. Finally, the APPS tab icons launch the start windows of the various toolboxes that are included in your MATLAB installation. For virtually any icon in any of the tabs, if you move the mouse pointer over it, you will see a capsule description of its function.

Many of the icons on these tabs also have keyboard shortcuts, listed to the right of their description. Some of these shortcuts depend on your operating system, and we generally will not mention them. Nonetheless, you may find it useful to note and use the keyboard shortcuts for Toolstrip items that you frequently use.

A small tool bar survives in the latest version of the Desktop. It is just below the Toolstrip and contains buttons and a window in which you can control the current folder and path (see the section below "The Current Folder and Search Path").

Each of the windows in the Desktop contains a small button ⦿ in the upper right-hand corner. If you click on it, you will see a menu of choices that includes one labeled ×, which allows you to close the window, while the curved arrow will "undock" the window from the Desktop. You can return it to the Desktop by selecting **Dock** in the same fashion on the undocked window.

The Workspace

In Chapter 2, we introduced the commands **clear** and **whos**, which can be used to keep track of the variables you have defined in your MATLAB session. The complete collection of defined variables is referred to as the Workspace, which you can view using the Workspace Browser. You can make this browser appear by typing **workspace** or by bringing up the default layout of the MATLAB Desktop. The Workspace Browser contains a list of the current variables and their values. If you double-click on a variable, its contents will appear in a new window called the *Variables Editor*, which you can use to edit individual entries in a vector or matrix. (You can also right-click and choose an appropriate option, or employ the **Open Variable** button in the VARIABLE section of the Toolstrip. Alternatively, the command **openvar('v')** also will open the Variables Editor and display the value of the variable **v**.) You can remove a variable from the Workspace by selecting it in the Workspace Browser and choosing **Edit:Delete**.

If you need to interrupt a session and don't want to be forced to recompute everything later, then you can save the current Workspace with **save**. For example, typing **save myfile** saves the values of all currently defined variables in a file called myfile.mat. To save only the values of the variables **X** and **Y**, type:

```
>> save myfile X Y
```

When you start a new session and want to recover the values of those variables, use **load**. For example, typing **load myfile** restores the values of all the variables stored in the file myfile.mat.

✓   By default, variables are stored in a binary format that is specific to MATLAB, but you can also **save** or **load** data as ASCII text. For details, see the online help for these commands. This feature is useful for exchanging data with other programs.

The Current Folder and Search Path

New files that you create from within MATLAB will be stored in your *current folder*. The name of this folder is displayed in the MATLAB Desktop tool bar (see above), and the files and subfolders it contains are listed in the Current Folder Browser. You can also display the name of the current folder by typing **pwd** ("print working directory") in the Command Window, and can get a list of the folder's contents by typing **dir** or **ls**.

✓   If you click on a folder in the Current Folder Browser, the files that it contains are listed. If you then click on a file, you can see its section or cell headings in the Details Window below. (See "Cell Mode" on page 30 later in this chapter.)

✓   The term *folder* is now more common than *directory*; for a computer file system, they mean the same thing. We will use "folder" because MATLAB uses this term in its documentation. MATLAB formerly used the term "directory" and you might see that term in old MATLAB documentation or in other resource material.

You may want to change the current folder from its default location, or you may want to maintain different folders for different projects. You can change the current folder in MATLAB by using the command **cd**, the Current Folder Browser, or the Current Folder box on the Desktop tool bar. You can type the folder name into this box and type ENTER, select a folder you have used before by clicking on the down-arrow at the right of the box, or browse for a folder by clicking on the "Browse for folder" icon to the left of the box.

✓   The command **cd** stands for "change directory." Although absolutely standard, it remains a vestige of the time when *directory*, rather than *folder*, was the common term.

On a *Windows* computer, the default current folder is likely to be:

$$C:\backslash Users\backslash [User\ Name]\backslash Documents\backslash MATLAB$$

You can create a new folder, say ProjectA, within it by typing **mkdir ProjectA**. You can also right-click in the Current Folder Browser and select **New Folder**. Then type **cd ProjectA** or double-click on it in the Current Folder Browser to make it your current folder. You will then be able to read and write files in this folder in your current MATLAB session.

If you need only to be able to read files from a certain folder, an alternative to making it your current folder is to add it to the *path* of folders that MATLAB searches to find files. The current folder and the folders in your path are the only places MATLAB searches for files, unless you explicitly type the folder name as part of the file name. To see the current path, type **path**. To add the folder ProjectA to your path, type:

>> addpath C:\Users\[User Name]\Documents\MATLAB\ProjectA

When you add a folder to the path, the files it contains remain available for the rest of your session regardless of whether you subsequently add another folder to the path or change the current folder. The potential disadvantage of this approach is that you must be careful when naming files. When MATLAB searches for files, it uses the first file with the correct name that it finds in the path list, starting with the current folder. If you use the same name for different files in different folders in your path, you can run into problems.

You can also control the MATLAB search path from the "Set Path" tool. To open this tool, type **pathtool**, or select **Set Path** in the ENVIRONMENT section of the Toolstrip. The Set Path tool consists of a panel, with a list of folders in the current path, and several buttons that allow you to add, remove, and re-order the folders in your path.

✓ If you have many toolboxes installed, path searches can be slow, especially with **lookfor**. Removing the toolboxes you are not currently using from the MATLAB path is one way to speed up execution.

☞ *Changes you make to the current folder and path are not saved from one MATLAB session to the next. At the end of the* Script M-files *section below, we describe how to change these and other items automatically each time you start MATLAB.*

The Command History Window

The Command History Window contains a running history of the commands that you type into the Command Window. It is useful in two ways. First, it lets you see at a quick glance a record of the commands that you have entered previously. Second, it can save you some typing time. If you double-click on an entry in the Command History Window, then it will be executed immediately in the Command Window. However, often you will want to edit a previous command before executing it. If you right-click (that is, click with the right mouse button) on an entry in the Command History Window, it becomes highlighted and a menu of options appears. You can select **Copy**, then right-click in the Command Window and select **Paste**, whereupon the command you selected will appear at the command prompt and be ready for editing. As we described in *Recovering from Problems* in the previous chapter, you can also type the UP- and DOWN-ARROW keys in the Command Window to scroll through the commands that you have used recently. Then when you locate the correct command line, you can use the LEFT- and RIGHT-ARROW keys to move around in the command line, deleting and inserting changes as necessary, and then press ENTER to tell MATLAB to evaluate the modified command.

For example, you might want to calculate to 15 digits the values of $\sin(0.1)/0.1$, $\sin(0.01)/0.01$, and $\sin(0.001)/0.001$. Here is a first try at a solution, together with the response that MATLAB displays in the Command Window:

```
>> x = [0.1, 0.01, 0.001];
>> y = sin(x)/x

y =
    0.9984
```

This is not the intended result; only one number is displayed instead of three and, furthermore, it is not $\sin(0.1)/0.1$, which is actually 0.9983. Remember that to divide two vectors element-by-element, you must type **./** rather than **/**. (Typing only **/**

"solves" the equation **y*x = sin(x)** for **y** in the "least-square" sense; type **help slash** for more information.) Another problem is that only five digits are displayed, not 15. To correct these problems, first type **format long**. Then type UP-ARROW twice to redisplay the command defining **y**, and type RIGHT-ARROW twice to move the cursor between the **)** and **/**. Finally, type **.** and ENTER:

```
>> y = sin(x)./x

y =
    0.998334166468282    0.999983333416666    0.999999833333342
```

M-Files

M-files allow you to save multiple MATLAB commands in a file and then run them with a single command or mouse click. While you may have solved the simple problem above correctly on the first try, more complicated problems generally require some trial and error – running, editing, and re-running a series of commands several times. While the Command History Window can be useful during the first stages of this process, eventually you will find it more efficient to use M-files. M-files also allow you to share your solution to a problem with other MATLAB users, and to format your results for others to read. There are two different kinds of M-files: script M-files and function M-files. We shall illustrate the use of both types of M-files as we present different solutions to the problem described above.

M-files are ordinary text files containing MATLAB commands. You can create and modify them using any text editor or word processor that is capable of saving files as plain ASCII text. (Such text editors include *Notepad* and *WordPad* in Windows, and **emacs** and **vi** in UNIX.) More conveniently, you can use the built-in *Editor/Debugger*, which you can start by typing **edit**, either by itself (to edit a new file) or followed by the name of an existing M-file in the current folder. You can also use any of the four icons (**New Script**, **New**, **Open**, or **Find Files**) in the FILE section of Toolstrip to start the Editor/Debugger, either to create a new M-file or to open an existing M-file. Double-clicking on an M-file in the Current Folder Browser will also open it in the Editor/Debugger.

The Editor/Debugger actually opens as the first of three tabs in a new Toolstrip in a special MATLAB Editor window. The other two tabs are labelled PUBLISH and VIEW. We'll discuss these later in Chapter 7, although we'll have a few words to say now about the buttons in the PUBLISH tab in "Presenting Your Results" later in this chapter.

Script M-Files

A script M-file contains a sequence of MATLAB commands to be run in order. We now show how to construct a script M-file to solve the mathematical problem described earlier. Create a file containing the following lines:

```
format long
x = [0.1, 0.01, 0.001];
y = sin(x)./x
```

We will assume that you have saved this file with the name `task1.m` in your current folder, or in some folder in your path. You can name the file any way you like (subject to the usual naming restrictions on your operating system), but the ".m" suffix is mandatory.

You can tell MATLAB to run (or *execute*) this script by typing **task1** in the Command Window. (You must *not* type the ".m" extension here; MATLAB automatically adds it when searching for M-files.) The output – but not the commands that produce them – will be displayed in the Command Window. Now the sequence of commands can easily be changed by modifying the M-file `task1.m`. For example, if you also wish to calculate $\sin(0.0001)/0.0001$, you can modify the M-file to read

```
format long
x = [0.1, 0.01, 0.001, 0.0001];
y = sin(x)./x
```

and then run the modified script by again typing **task1**. Be sure to save your changes to `task1.m` first; otherwise, MATLAB will not recognize them.

✓   Any variables that are set by running a script M-file will persist exactly as if you had typed them into the Command Window directly. For example, the program above will cause all future numerical output to be displayed with 15 digits. In order to revert to five-digit format, you would have to type **format short**.

Adding Comments   It is worthwhile to include comments in M-files. These comments might explain what is being done in the calculation, or might interpret the results of the calculation. In MATLAB, the percent sign (**%**) begins a comment; the rest of the line is not executed by MATLAB. (The Editor/Debugger colors comments green to help distinguish them from commands, which appear in black.) Here is our new version of `task1.m` with a few comments added:

```
format long    % turn on 15 digit display
x = [0.1, 0.01, 0.001];
y = sin(x)./x
% These values illustrate the fact that the limit of
% sin(x)/x as x approaches 0 is 1.
```

Notice that a multi-line comment requires a percent sign at the beginning of each line.

Cell Mode   An important feature in MATLAB allows you to divide a script M-file into subunits called *cells*. This is especially useful if your M-file is long or if you are going to *publish* it, as explained below in *Publishing an M-file*. To start a new cell, insert a comment line (which will serve as the "title" of the cell that follows) starting with a double percent sign **%%** followed by a space. You can type in the double percent and a space directly in the file or you can use the buttons in the Editor's Toolstrip. To do the latter, use either the **Insert** button in the EDITOR tab or the **Section** button in the PUBLISH tab. If you want to give the cell a name, then following the space, type its name. You can enter more cells at any time by entering a double percent sign followed by a space (and then a cell name if you wish). Then when you click somewhere in the M-file, the cell that contains that location will be highlighted in pale yellow. You can evaluate that cell by then clicking on one of the buttons in the RUN section. This can be a big help if you've made a change in just one cell and do not want to run the whole script all over again. You should explore the different buttons in the RUN section, e.g., **Run Section** or **Run and Advance**.

Initializing Script M-files   In order for the results of a script M-file to be reproducible, the script should be self-contained, unaffected by other variables that you might have defined elsewhere in the MATLAB session, and uncorrupted by leftover graphics. For example, if you define a variable named **sin** in the Command Window and then run the script `task1.m` above, you will get an error message because **sin** now represents a variable rather than the usual built-in function. With this in mind, you can include the line **clear all** at the beginning of the script to ensure that previous definitions of variables do not affect the results. You can also include **close all** at the beginning of a script M-file that creates graphics, to close all figure windows and start with a clean slate.

As mentioned above, the commands in a script M-file will not automatically be displayed in the Command Window. If you want the commands to be displayed along with the results, add the command **echo on** to the beginning of the script file (it is then a good idea to add **echo off** to the end of the script). Any comments in the M-file will then be echoed too. When running a long script M-file, echoing is useful to keep track of which outputs go with which inputs.

Here is a more carefully commented version of `task1.m` that will display input as well as output:

```
clear all    % remove old variable definitions
echo on      % display the input in the command window
format long    % turn on 15 digit display

x = [0.1, 0.01, 0.001];    % define the x values
y = sin(x)./x    % compute the desired quotients

% These values illustrate the fact that the limit of
% sin(x)/x as x approaches 0 is equal to 1.

echo off
```

Startup M-File   When MATLAB starts, it searches the default path (which includes the default current folder) for a script M-file called `startup.m`. If you create such a file, the commands it contains will be run each time MATLAB starts. You can use this file to save customizations that do not normally persist from one session to the next, such as changes to the current folder or path. (In addition to the commands **cd** and **addpath** described above, you can remove folders from the path with **rmpath**.)

Function M-Files

Function M-files, unlike script M-files, allow you to specify input values when you run them from the MATLAB command line or from another M-file. As we described in the previous chapter, you can also use the anonymous-function (**@**) syntax to define your own functions on the command line. However, this method allows only a one-line function definition, so M-files are necessary to define more complicated functions. Like a script M-file, a function M-file is a plain text file that should reside in your current folder or elsewhere in your MATLAB path.

Let us return to the problem described above, where we computed some values of $\sin(x)/x$ with $x = 10^{-b}$ for several values of b. Suppose, in addition, that you want to find the smallest value of b for which $\sin(10^{-b})/(10^{-b})$ and 1 agree to 15 digits. Here is a function M-file called `sinelimit.m` designed to explore this question:

```
function y = sinelimit(c)
% SINELIMIT computes sin(x)/x for x = 10^(-b),
% where b = 1, ..., c.
format long
b = 1:c;
x = 10.^(-b);
y = (sin(x)./x)';
```

The first line of the file starts with **function**, which identifies the file as a function M-file. (The Editor/Debugger colors this special word blue.) The first line of the M-file specifies the name of the function and describes both its input arguments (or parameters) and its output values. In this example, the function is called **sinelimit**. The file name (without the .m extension) and the function name should match. When you create this new function M-file in an untitled editor window and select **Save**, the Editor/Debugger knows to call the file `sinelimit.m`. The function in our example takes one input, which is called **c** inside the M-file. It also returns one output, which is the value of **y** that results when the function finishes executing.

It is good practice to follow the first line of a function M-file with one or more comment lines explaining what the M-file does. If you do, **help** will automatically retrieve this information. For example:

```
>> help sinelimit
```

```
  SINELIMIT computes sin(x)/x for x = 10^(-b),
  where b = 1, ..., c.
```

The remaining lines of the M-file define the function. In this example, **b** is defined to be a row vector consisting of the integers from **1** to **c**, then **x** is computed from **b**, and finally **y** is determined from **x**.

☞   *The variables used in a function M-file, like* **b**, **x**, *and* **y** *in* `sinelimit.m`, *are* local *variables. This means that, unlike the variables that are defined in a script M-file, these variables are completely unrelated to any variables with the same names that you may have used in the Command Window, and MATLAB does not remember their values after the function M-file has been executed. For further information, see* Variables in Function M-Files *in Chapter 4.*

Notice that the output of the lines defining **b**, **x**, and **y** is suppressed with a semi-colon. Since these variables are only used internally by the M-file, it would be misleading to see output about them in the Command Window. While displaying the output of intermediate calculations can be useful for debugging, in general you should suppress all output in a function M-file.

Here is an example that shows how to use the function **sinelimit**:

```
>> sinelimit(5)
```

```
ans =
    0.998334166468282
    0.999983333416666
    0.999999833333342
    0.999999998333333
    0.999999999983333
```

None of the values of b from 1 to 5 yields the desired answer, 1, to 15 digits. Judging from the output, you can expect to find the answer to the question we posed above by typing **sinelimit(10)**. Try it!

Loops

A *loop* specifies that a command or group of commands should be repeated several times. The easiest way to create a loop is to use a **for** statement. Here is a simple example that computes and displays $10! = 10 \cdot 9 \cdot 8 \cdots 2 \cdot 1$:

```
f = 1;
for n = 2:10
    f = f*n;
end
f
```

The loop begins with the **for** statement and ends with the **end** statement. The command between those statements is executed a total of nine times, once for each value of **n** from 2 to 10. We used a semicolon to suppress intermediate output within the loop. In order to see the final output, we then needed to type **f** after the end of the loop. Without the semicolon, MATLAB would display each of the intermediate values $2!, 3!, \ldots$.

The Editor/Debugger automatically colors the commands **for** and **end** blue. It is not necessary, but improves readability, if you indent the commands in between (as we did above); the Editor/Debugger does this automatically. If you type **for** in the Command Window, MATLAB does not give you a new prompt (**>>**) until you enter an **end** command, at which time MATLAB will evaluate the entire loop and display a new prompt.

✓ If you use a loop in a script M-file with **echo on** in effect, the commands will be echoed every time through the loop. You can avoid this by inserting the command **echo off** just before the **end** statement and inserting **echo on** just afterward; then each command in the loop (except **end**) will be echoed once.

Presenting Your Results

Sometimes you may want to show other people the results of a script M-file that you have created. For a polished presentation, you can import your results into another program, such as a word processor, or use the PUBLISH tab in the Editor/Debugger to convert your results to a format such as PDF or HTML, or (on a *Windows* computer) use an M-book. To share your results more informally, you can give someone else your M-file, assuming that person has a copy of MATLAB on which to run it, or you can provide the output you obtained in the form of a *diary* file. We now discuss these different approaches.

✓   You can greatly enhance the readability of your M-file by including frequent comments. Your comments should explain what is being calculated, so the reader can understand your procedures and strategies. Once you've done the calculations, you can also add comments that interpret the results.

Publishing an M-File

MATLAB comes with a very convenient *publish* feature, which enables the user to convert a script M-file into a readable document. You should use this feature in combination with cells; see *Cell Mode* above. The default output formats, which we have found work best, are HTML (i.e., a web page) and PDF, but you can also publish to a *Word* document, a *PowerPoint* presentation (on a *Windows* computer), or to a *LATEX* document. You can use the command **publish** (see its online help), but the most convenient way to invoke this feature is to click the green triangle icon in the PUBLISH section of the PUBLISH tab.

✓   In the PUBLISH section, there is also a drop down menu that contains an item **Edit Publishing Options...**, which allows you to choose the format in which to publish.

Publishing an M-file reproduces the M-file along with its text and graphics output. Each cell of the M-file appears as a separate section, under the heading taken from the appropriate line starting with a double percent sign **%%** followed by a space. Any comment lines that immediately follow this line appear as formatted text in the output, provided that they start with a percent sign and a space, and that there are no intervening blank lines. (You can select several formats, such as bulleted lists, using various buttons in the INSERT INLINE MARKUP and INSERT BLOCK MARKUP sections of the Editor's Toolstrip.) Next, all of the MATLAB input lines and any remaining comments in the cell are reproduced without any formatting, followed by any text output (in gray), followed by graphics. (It is not necessary to use **echo** when publishing an M-file; the input and comments appear in the published document, unless you set the "Include code" option to false.)

✓   A line that starts with a double percent sign but has no following text starts a new cell but does not produce a section heading in the published output. Starting a new cell in this way allows you to insert formatted text that discusses the output from the previous cell, without starting a new section. Perhaps more importantly, it allows you to keep input and output close to each other in a section with many input lines. If you don't break up such a section into smaller cells, all of its input will appear before all of its output, making it hard to match output to input.

Here is another version of the script M-file `task1.m` discussed above that is suitable for publishing:

```
%% Sample Script M-file
% This script computes sin(x)/x for x = 0.1, 0.01, 0.001.

clear all   % remove old variable definitions
format long   % turn on 15 digit display

x = [0.1, 0.01, 0.001];   % define the x values
y = sin(x)./x   % compute the desired quotients

%%
% These values illustrate the fact that the limit of
% sin(x)/x as x approaches 0 is equal to 1.
```

If your M-file puts two successive figures in the same figure window, then only the last figure will be written to the published document, unless the figures appear in different cells. So you should either start a new cell each time you create a new figure, or else use **figure** to open separate windows for each figure produced by a given cell.

☞   *See Chapter 7 for a more thorough discussion of publishing. See Chapter 5 for more about figure windows.*

Diary Files

MATLAB has a much more primitive ancestor of the **publish** command called **diary**. It saves all of the text output (and, with **echo on**, the input) from a session to a file. However, if your code generates any graphics output, **diary** does not capture it. Truthfully, only if you were stuck in an ASCII world and could not access HTML or PDF, etc., would you utilize **diary**. But if that is your situation, then here is a way to use **diary** to automatically save the text input and output from a script M-file to a file each time you run it. At the beginning of the M-file, say `task1.m`, you can include the commands:

```
delete task1.txt
diary task1.txt
echo on
```

The script M-file should then end with the commands:

```
echo off
diary off
```

The first **diary** command causes all subsequent input to and output from the Command Window to be copied into the specified file – in this case, `task1.txt`. The diary file `task1.txt` is a plain text file that is suitable for printing or importing into another program.

By using **delete** at the beginning of the M-file, you ensure that the file only contains the output from the most recent run of the script. If you omit the **delete** command, then **diary** will add any new output to the end of an existing file, and the file `task1.txt` can end up containing the results of several runs of the M-file.

(Putting the **delete** command in the script will lead to a harmless warning message about a nonexistent file the first time you run the script.) You can also get extraneous output in a diary file if you type CTRL+C to halt a script containing a **diary** command. If you do so, you should then type **diary off** in the Command Window before proceeding.

Interactive M-Files

If you want someone else to run your script M-file and see the results, it is helpful to use the command **pause** to stop execution at various points: after each graph is produced, after important comments, and after critical places where your script generates numerical output. Each time MATLAB reaches a **pause** command, it waits until the user presses a key before proceeding. Of course, if the recipient of your M-file is not familiar with **pause**, then you should include appropriate guidance at such points; for example, with **echo on** in effect you could type:

```
pause    % Press any key to continue...
```

If you are not using **echo on**, then you can use the command **disp** to display an appropriate message before pausing. Another command that may be useful is **input**, which both displays a message and waits for the user to enter a number or other MATLAB expression; see the online help.

✓   These interactive commands can be an annoyance if, for example, you want to publish the M-file. If your recipient is familiar with cell mode in the Editor/Debugger, then an alternative to using **pause** is to separate the M-file into cells, and let your recipient run the file one cell at a time.

☞   *MATLAB also has commands and tools for developing a graphical user interface (GUI); see Chapter 10. This approach allows your recipient to interact with your M-file in a separate window that you design, rather than in the Command Window.*

Wrapping Long Input and Output Lines

Sometimes you may need to type, either in the Command Window or in an M-file, a command that is too long to fit on one line. If so, when you get near the end of a line you can type **...** (that is, three successive periods) followed by ENTER, and continue the command on the next line. If you do this in the Command Window, you will not see a command prompt on the new line.

Symbolic   Numerical output is formatted by MATLAB to fit your screen, but symbolic expressions generally appear on one line no matter how long they are. If **s** is a symbolic expression, then typing **pretty(s)** displays **s** in *pretty print* format, which uses multiple lines on your screen to imitate written mathematics. The result is often easier to read than the default format. An important feature of **pretty** is that it *wraps* long expressions to fit within the margins (80 characters wide) of a standard-sized window. If your symbolic output is long enough to extend past the right-hand edge of your window, it may be truncated when you print your output, so you should use **pretty** to make the entire expression visible in your printed output.

Printing and Saving Graphics

As indicated in Chapters 1 and 2, graphics appear in a separate window. You can save the current graph by selecting **File:Save As...** in the figure window. The default file format (.fig) is specific to MATLAB and is not understood by most other programs; however, several other formats are available. You can also print the current figure by selecting **File:Print....** Alternatively, typing **print** causes the figure in the current figure window to be printed on your default printer. Since you probably don't want to print the figure every time you run a script, you should not include a bare **print** command in an M-file. Instead, you should use a form of **print** that sends the output to a file. It is also helpful to give informative titles to your figures. For example:

```
close all    % remove old figure windows

% Graph sin(x) from 0 to 2*pi:
x = 2*pi*(0:0.01:1);
plot(x, sin(x))

title('Figure A: Sine Curve')    % title the figure
print -deps figureA    % store the graph in figureA.eps
```

The form of **print** used in this script causes the current figure to be written in *Encapsulated PostScript*® format to a file in the current folder, called figureA.eps. This file can be printed later (on a PostScript printer), or it can be imported into another program that can read PostScript format. If you intend to paste MATLAB graphics into a web page, it is better to save the figure as a png file with **print -dpng**. (You can also select **Save As...** from the **File** menu of the figure window and under "Save as type:" select "Portable Network Graphics file (*.png)".) See the online help for **print** for other formats available – e.g., "Portable Document Format (*.pdf)".

As a final example involving graphics, let's consider the problem of graphing the functions $\sin(x)$, $\sin(2x)$, and $\sin(3x)$ on the same set of axes. This is a typical example; we often want to plot several similar curves whose equations depend on a parameter. Here is a script M-file solution to the problem:

```
x = 2*pi*(0:0.01:1);    % define the x values
figure, hold on    % open a new figure window with overlay

% Run a loop to plot three sine curves:
for c = 1:3
   plot(x, sin(c*x))
end

hold off, axis([0, 2*pi, -1, 1])    % adjust the x axis
title('Several Sine Curves')    % title the figure
```

The result is shown in Figure 3.1.

Let's analyze this solution. We start by defining a grid of 101 evenly spaced x-values from 0 to 2π. Then **figure** opens a new figure window, and **hold on** lets

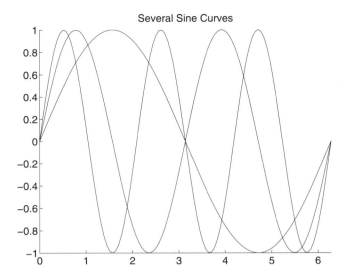

Figure 3.1. Sine Curves.

MATLAB know that we want to draw several curves on the same set of axes. Rather than type three **plot** commands, we use a **for** loop, as described above. Then we use **hold off** to release the figure window, make the graph look a little nicer by changing the horizontal axis from the range MATLAB chooses to the actual range of x, and **title** the figure.

M-Books

Another sophisticated way of presenting MATLAB output, without having to produce an M-file first, is in a Microsoft *Word* document, incorporating text, MATLAB commands, and graphics. This allows you more control over formatting than **publish** affords. A simple first approach is to prepare a *Word* document with explanatory comments, and to paste in your MATLAB commands (you can do this in another font) or include your M-files using **File:Text:Text from File** in *Word*. Finally, you can paste in the graphics by selecting **Insert:Picture from File** in *Word*. You should have first saved the graphics in a common format such as PNG, TIFF, or EPS.

A more direct approach is to enable M-books on your computer. An M-book is a *Word* document with embedded executable MATLAB code (that runs as a *Word* macro, via the intermediary of *Visual Basic*). You can launch an M-book by typing **notebook** in the Command Window, or by starting *Word*, choosing **New...** from its **File** menu, and selecting m-book as the Document template. If the file m-book.dot does not already exist on your computer, you need to *enable M-books* first. This is done by typing

```
>> notebook -setup
```

in the MATLAB Command Window and following the instructions.

⇨ **For M-books to run properly, you must enable execution of the necessary macros in *Word*. The safest way to do this is to see whether you get a security warning dialog box when you open an M-book and, if so, select "Always trust macros from this source". This will maintain your security level at "high" for other macros. Alternatively, you can set your security level to "medium" for all macros in the dialog box you get by selecting Tools:Macros:Security... in *Word*.**

Once you have successfully launched an M-book, it will behave just like any other *Word* document, except for the **Notebook** menu, which is found under **Add-Ins**, at the top. If you type a MATLAB command and hit CTRL+ENTER, or else highlight the command with the mouse and select **Evaluate Cell** in the **Notebook** menu, MATLAB will evaluate the command and send the output back to *Word*. For ease in reading, *Word* typesets "input cells" (MATLAB input) in green Courier bold and "output cells" (MATLAB output) in blue Courier type. You have an option (which you can adjust with the **Toggle Graph Output for Cell** item in the **Notebook** menu) of having figure windows appear separately, having them appear in the M-book, or both.

In one respect, M-books behave like M-files; you can modify them and run them again and again. If you find you mistyped a command or want to change a command, you can simply go back to the appropriate input cell, change it, and then re-evaluate it. The new output will replace the old. Keep in mind, though, that the output will reflect the order in which you evaluate the cells, not the order in which they appear in the M-book. When preparing your final document, you should re-run the entire M-book by selecting **Notebook:Evaluate MATLAB Notebook** and make sure that the commands produce the output you desire when they are run in order.

Fine-Tuning Your M-Files

Script M-files allow you to refine your strategy for solving a problem without starting from scratch each time. Nonetheless, debugging a lengthly M-file can be time-consuming. Here are some general tips for diagnosing errors (and avoiding them in the first place).

☞ *We will discuss features of the Editor/Debugger and more advanced MATLAB debugging commands in the section* Debugging *in Chapter 6 and in the section* Debugging Techniques *in Chapter 12.*

- Use publishing with frequent cell dividers, or use **echo on** near the beginning of an M-file, so that you can see "cause" as well as "effect."

- Use cell mode in the Editor/Debugger, or make liberal use of **pause**, so that you can see the results of one part of an M-file at a time.

- Use the semicolon to suppress unneeded output, especially large vectors and arrays that you define.

- If you are producing graphics with **hold on**, don't forget to type **hold off** before starting a new figure, so that the next figure is not mixed with the old one.

- Also, when starting a new figure, insert a cell divider or **pause** command, or use **figure** to open a new window, so that the next graphics command will not obliterate the current one before you have a chance to see it.

- Do not include bare **print** commands in your M-files. Instead, print to a file.

- The command **keyboard** is like **pause**, but is more useful for debugging. If you have the line **keyboard** in your M-file, then when MATLAB reaches it, execution of your script is interrupted, and a new prompt appears with the letter **K** before it. At this point you can type any normal MATLAB command. This is useful if you want to examine or reset some variables in the middle of running a script. To resume execution, type the command **return**.

- Finally, remember that you can stop a running M-file by typing CTRL+C. This is useful if a calculation is taking too long, if MATLAB is spewing out undesired output, or if, at a **pause** command, you realize that you want to stop execution completely.

Practice Set A

Algebra and Arithmetic

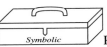

Problems 3–8 require the Symbolic Math Toolbox. The others do not.

1. Compute the following quantities:

    (a) $1111 - 345$.

    (b) e^{14} and 382801π to 15 digits each. Which is bigger?

    (c) the fractions $2709/1024$, $10583/4000$, and $2024/765$. Which of these is the best approximation to $\sqrt{7}$?

2. Compute to 15 digits the following quantities:

    (a) $\cosh(0.1)$.

    (b) $\ln(2)$. (*Hint*: the natural logarithm in MATLAB is called **log**, not **ln**.)

    (c) $\arctan(1/2)$. (*Hint*: the inverse tangent function in MATLAB is called **atan**, not **arctan**.)

3. Solve (symbolically) the system of linear equations

$$3x + 4y + 5z = 2$$
$$2x - 3y + 7z = -1$$
$$x - 6y + z = 3.$$

    Check your answer using matrix multiplication.

4. Try to solve (symbolically) the system of linear equations

$$3x - 9y + 8z = 2$$
$$2x - 3y + 7z = -1$$
$$x - 6y + z = 3.$$

    What happens? Can you see why? Again check your answer using matrix multiplication. Is the answer "correct"?

5. Factor the polynomial $x^4 - y^4$.

6. Use **simplify** or **simple** to simplify the following expressions:

    (a) $1/\left(1 + 1/\left(1 + \frac{1}{x}\right)\right)$,

41

    (b) $\cos^2 x - \sin^2 x$.

7. Compute 3^{301}, both as an approximate floating-point number and as an exact integer (written in usual decimal notation).

8. Use either **solve** or **fzero**, as appropriate, to solve the following equations:

    (a) $67x + 32 = 0$ (exact solution).

    (b) $67x + 32 = 0$ (numerical solution to 15 places).

    (c) $x^3 + px + q = 0$. (Solve for x in terms of p and q.)

    (d) $e^x = 8x - 4$ (*all* real solutions). It helps to draw a picture first.

9. Use **plot** and/or **ezplot**, as appropriate, to graph the following functions:

    (a) $y = x^3 - x$ for $-4 \le x \le 4$.

    (b) $y = \sin(1/x^2)$ for $-2 \le x \le 2$. Try this with both **plot** and **ezplot**. Are both results "correct"? If you use **plot**, be sure to plot enough points.

    (c) $y = \tan(x/2)$ for $-\pi \le x \le \pi, -10 \le y \le 10$. (*Hint*: first draw the plot; then use **axis**.)

    (d) $y = e^{-x^2/2}$ and $y = x^4 - x^2$ for $-1.5 \le x \le 1.5$ (on the same set of axes).

10. Plot the functions x^4 and 2^x on the same graph and determine how many times their graphs intersect. (*Hint*: you will probably have to make several plots, using intervals of various sizes, in order to find all the intersection points.) Now find the approximate values of the points of intersection using **fzero**.

Chapter 4

Beyond the Basics

In this chapter, we describe some of the finer points of MATLAB and review in more detail some of the concepts introduced in Chapter 2. We explore enough of MATLAB's internal structure to improve your ability to work with complicated functions, expressions, and commands. At the end of this chapter, we introduce some of the MATLAB commands for doing calculus.

Suppressing Output

Some MATLAB commands produce output that is superfluous. For example, when you assign a value to a variable, MATLAB echoes the value. But, as we mentioned briefly in Chapter 2, you can suppress the output of a command by putting a *semicolon* after the command. Here is an example:

```
>> syms x
>> y = x + 7

y =
x + 7

>> z = x + 7;
>> z

z =
x + 7
```

The semicolon does not affect the way MATLAB processes the command internally, as you can see from its response to the command **z**.

You can also use semicolons to separate a series of commands when you are interested only in the output of the final command (several examples appear later in the chapter). As we noted in Chapter 2, commas can also be used to separate commands – without suppressing output. If you use a semicolon after a graphics command, it will not suppress the graphic.

⇨ **The most common use of the semicolon is to suppress the printing of a long vector or big matrix.**

Another object that you may want to suppress is MATLAB's label for the output of a command. The command **disp** is designed to achieve that; typing **disp(x)** will print the value of the variable **x** without printing the label and the equal sign. So,

```
>> x = 7; disp(x)
    7
```

or:

```
>> disp(solve('x + tan(y) = 5', 'y'))
-atan(x - 5)
```

Data Classes

Every variable you define in MATLAB, as well as every input to, and output from, a command, is an *array* of data belonging to a particular *class*. In this book, we use primarily the following types of data: floating-point numbers, symbolic expressions, character strings, and function handles. We introduced each of these types in Chapter 2. An alternative to function handles, which you will sometimes find in older MATLAB code, involves inline functions. In Table 4.1, we list for each type of data its class (as given by **whos**) and how you can create it.

Table 4.1. MATLAB Data Classes.

Type of data	Class	Created by
Floating-point	`double`	Typing a number
Symbolic	`sym`	Using **sym** or **syms**
Character string	`char`	Typing a string inside single quotes
Function handle	`function_handle`	Using **@**
Inline function	`inline`	Using **inline**

You can think of an array as a two-dimensional grid of data. A single number (or symbolic expression) is regarded by MATLAB as a 1×1 array, sometimes called a *scalar*. A $1 \times n$ array is called a *row vector*, and an $m \times 1$ array is called a *column vector*. (A string is actually a row vector of characters.) An $m \times n$ array of numbers is called a *matrix*; see *More on Matrices* below. You can see the class and array size of every variable you have defined by looking in the Workspace Browser or typing **whos** (see *Managing Variables* in Chapter 2). The set of variable definitions shown by **whos** is called your *Workspace*.

In order to use MATLAB commands effectively, you must pay close attention to the class of data each command accepts as input and returns as output. The input to a command consists of one or more arguments separated by commas; some arguments are optional. Some commands, such as **whos**, do not require any input. The help text (see *Online Help* in Chapter 2) for each command usually states what classes of inputs the command expects as well as what class of output it returns.

✓   When you type a pair of words, like **hold on**, MATLAB interprets the second word as a string argument to the command given by the first word; thus, **hold on** is equivalent to **hold('on')**.

Many commands allow more than one class of input, and sometimes the same name may refer to commands in more than one toolbox (MATLAB, Simulink, the

Symbolic Math Toolbox, the Statistics Toolbox, etc.). This flexibility can be a convenience in some cases and a pitfall in others. Suppose that you have already defined **a = 10**, **b = 5**, and now you attempt to factor the expression $a^2 - b^2$, forgetting your previous definitions and that you have to declare the variables symbolic:

```
>> factor(a^2 - b^2)

ans =
   3     5     5
```

A more bizarre example of this would be if you defined **b = 'b'** and attempted to factor b^2:

```
>> factor(b^2)

ans =
   2     2     7     7     7     7
```

The reason you don't get error messages in these cases is that **factor** is the name of a command that factors integers into prime numbers as well as factoring expressions. In the first case, since $a^2 - b^2 = 75 = 3 \cdot 5^2$, the numerical version of **factor** is applied. In the second case, MATLAB again applies the numerical version of **factor**, but to the number 98^2. The reason is that in this case, **b** is of class char, and has been defined equal to **'b'**. This string, in turn, is associated by MATLAB to the numerical code 98, as you can see by typing **double('b')**. The output is clearly not what you intended, but in the course of a complicated series of commands, be careful not to be fooled by such unintended output.

✓ Note that typing **help factor** shows you only the help text for the numerical version of the command, but gives a cross-reference to the symbolic version at the bottom. If you want to see the help text for the symbolic version instead, type **help sym/factor**. Functions such as **factor** with more than one version are called *overloaded*.

Sometimes you need to convert one data class into another in order to prepare the output of one command to serve as the input for another. For example, we have used **double** to convert symbolic expressions (or even strings) to floating-point numbers and **sym** to convert numbers or strings to symbolic expressions. The commands **num2str** and **str2num** convert between numbers and strings, while **char** converts a symbolic expression to a string. You can also use **vectorize** to convert a symbolic expression to a vectorized string; it adds a . before every *, /, and ^ in the expression.

String Manipulation

Often it is useful to concatenate two or more strings together. The simplest way to do this is to use MATLAB's vector notation, keeping in mind that a string is a "row vector" of characters. For example, typing **[string1, string2]** combines **string1** and **string2** into one string.

A first useful application of string concatenation is to define a string variable containing an expression that takes more than one line to type. (In most circumstances you can continue your MATLAB input onto the next line by typing **...** followed by

ENTER or RETURN, but this is not allowed in the middle of a string.) The solution is
to break the expression into smaller parts and concatenate them, as in:

```
>> eqn = ['left hand side of equation = ', ...
'right hand side of equation']

eqn =
left hand side of equation = right hand side of equation
```

Another use of string concatenation is to format output using **disp**, since **disp**
can take only a single input, but this input is allowed to be a string:

```
>> disp(['The value of 10! is ', num2str(factorial(10)), '.'])

The value of 10! is 3628800.
```

✓   Notice that string output, like symbolic output, is not indented.

Symbolic and Floating-Point Numbers

We mentioned above that you can convert between symbolic numbers and floating-
point numbers with **double** and **sym**. By default, numbers that you type are floating-
point. However, if you mix symbolic and floating-point numbers in an arithmetic
expression, the floating-point numbers are automatically converted into symbolic numbers. This explains why you can type **syms x** and then **x^2** without having to convert
2 into a symbolic number. Here is another example:

```
>> a = 1

a =
   1

>> b = a/sym(2)

b =
1/2
```

MATLAB was designed so that some floating-point numbers are restored to their
exact values when converted into symbolic numbers. Integers, rational numbers with
small numerators and denominators, square roots of small integers, the number π, and
certain combinations of these numbers are so restored. For example:

```
>> c = sqrt(3)

c =
   1.7321

>> sym(c)

ans =
sqrt(3)
```

Since it is difficult to predict when MATLAB will preserve exact values, it is best
to suppress the floating-point evaluation of a numerical argument to **sym** by enclosing
it in single quotes to make it a string, e.g., **sym('1 + sqrt(3)')**. We will see
below another way in which single quotes suppress evaluation.

Functions and Expressions

We have used the terms *expression* and *function* without carefully making a distinction between the two. Strictly speaking, if we define $f(x) = x^3 - 1$, then f (written without any particular input) is a function while $f(x)$ and $x^3 - 1$ are expressions involving the variable x. In mathematical discourse, we often blur this distinction by calling $f(x)$ or $x^3 - 1$ a function, but in MATLAB the difference between functions and expressions is important.

In MATLAB, an expression can belong to either the string or the symbolic class of data. Consider the following example:

```
>> f = 'x^3 - 1';
>> f(7)

ans =
1
```

This result may be puzzling if you are expecting **f** to act like a function. Since **f** is a string, **f(7)** denotes the seventh character in **f**, which is **1** (the spaces count). Recall that, like symbolic output, string output is not indented from the left-hand margin. This is a clue that the answer above is a string (consisting of one character), not a floating-point number. Typing **f(5)** would yield – and **f(-1)** would produce an error message.

You have learned two ways to define your own functions: using **@** to create an anonymous function (see Chapter 2) or using an M-file (see Chapter 3). Another option is to use **inline** to create an inline function, but, as noted in the online help, this option may be removed in future versions of MATLAB. Anonymous or inline functions are most useful for defining simple functions that can be expressed in one line and for turning the output of a symbolic command into a function. Function M-files are useful for defining functions that require several intermediate commands to compute the output. Most MATLAB commands are actually M-files, and you can peruse them for ideas to use in your own M-files – to see the M-file for, say, the command **mean**, you can enter **type mean**. See also *More about M-files* below.

Some commands, such as **ode45** (a numerical ODE (ordinary differential equation) solver, which we use later in Chapter 11), require their first argument to be a function – to be precise, either an anonymous function (for example **ode45(@(t, y) t + y, [0 2], 1)**), an inline function (as in **ode45(f, [0 2], 1)**), or a *function handle*, that is, the name of a built-in function or a function M-file preceded by the special symbol **@** (as in **ode45(@func, [0 2], 1)**). With or without quotes, typing a symbolic expression as the first input to **ode45** gives an error message. On the other hand, most symbolic commands require their first argument to be either a string or a symbolic expression, not a function.

An important difference between strings and symbolic expressions is that MATLAB automatically substitutes user-defined functions and variables into symbolic expressions, but not into strings. (This is another sense in which the single quotes you type around a string suppress evaluation.) For example, if you type

```
>> h = @(t) t^3; int(sym('h(t)'), 't')

ans =
int(h(t), t)
```

then the integral cannot be evaluated because, within a string, **h** is regarded as an unknown function. But if you type

```
>> syms t, int(h(t), t)

ans =
t^4/4
```

then the previous definition of **h** is substituted into the symbolic expression **h(t)** before the integration is performed.

Substitution

In Chapter 2, we described how to create an anonymous function from an expression. You can then plug numbers into that function, to make a graph or table of values for instance. But you can also substitute numerical values directly into an expression with **subs**. See the subsection *Substituting in Symbolic Expressions* in Chapter 2 for instructions.

More about M-Files

Files containing MATLAB statements are called M-files. There are two kinds of M-files: *function M-files*, which accept arguments and produce output, and *script M-files*, which execute a series of MATLAB statements. In Chapter 3, we created and used both types. In this section, we present additional information on M-files.

Variables in Script M-Files

When you execute a script M-file, the variables you use and define belong to your Workspace; i.e., they take on any values you assigned earlier in your MATLAB session, and they persist after the script has finished executing. Consider the following one-line script M-file, called `scriptex1.m`:

```
u = [1 2 3 4];
```

Typing **scriptex1** assigns the given vector to **u**, but displays no output. Now consider another script M-file, called `scriptex2.m`:

```
n = length(u)
```

If you have not previously defined **u**, then typing **scriptex2** will produce an error message. However, if you type **scriptex2** after running **scriptex1**, then the definition of **u** from the first script will be used in the second script and the proper output n = 4 will be displayed.

   If you don't want the output of a script M-file to depend on any earlier computations in your MATLAB session, put the line **clear all** near the beginning of the M-file, as we suggested in *Structuring Script M-files* in Chapter 3.

Variables in Function M-Files

The variables used in a function M-file are *local*, meaning that they are unaffected by, and have no effect on, the variables in your Workspace. Consider the following function M-file, called `sq.m`:

```
function z = sq(x)
% sq(x) returns the square of x.
z = x.^2;
```

Typing **sq(3)** produces the answer **9**, whether or not **x** or **z** is already defined in your Workspace. Running the M-file neither defines them, nor changes their definitions if they have been previously defined.

Structure of Function M-Files

The first line in a function M-file is called the *function definition line*; it specifies the function name, as well as the number and order of input and output arguments. Following the function definition line, there can be several comment lines that begin with a percent sign **%**. These lines are called *help text* and are displayed in response to the command **help**. In the M-file `sq.m` above, there is only one line of help text; it is displayed when you type **help sq**. The remaining lines constitute the *function body*; they contain the MATLAB statements that calculate the function values. In addition, there can be comment lines (lines beginning with **%**) anywhere in an M-file. All statements in a function M-file that normally produce output should end with a semicolon to suppress the output.

Function M-files can have multiple input and output arguments. Here is an example, called `polarcoordinates.m`, with two input and two output arguments:

```
function [r, theta] = polarcoordinates(x, y)
% polarcoordinates(x, y) returns the polar coordinates
% of the point with rectangular coordinates (x, y).
r = sqrt(x^2 + y^2);
theta = atan2(y,x);
```

If you type **polarcoordinates(3,4)**, only the first output argument is returned and stored in **ans**; in this case, the answer is **5**. To see both outputs, you must assign them to variables enclosed in square brackets:

```
>> [r, theta] = polarcoordinates(3,4)

r =
   5
theta =
   0.9273
```

By typing **r = polarcoordinates(3,4)**, you can assign the first output argument to the variable **r**, but you cannot get only the second output argument; in fact, typing **theta = polarcoordinates(3,4)** will still assign the first output, **5**, to **theta**.

Complex Arithmetic

MATLAB does most of its computations using *complex numbers*; i.e., numbers of the form $a + bi$, where $i = \sqrt{-1}$, and a and b are real numbers. The complex number i is represented as **i** in MATLAB. Although you may never have occasion to enter a complex number in a MATLAB session, MATLAB often produces an answer involving complex numbers. For example, many polynomials with real coefficients have complex roots:

```
>> syms x; solve(x^2 + 2*x + 2 == 0)

ans =
  - 1 + i
  - 1 - i
```

Both roots of this quadratic equation are complex numbers, expressed in terms of the number i. Some common functions also return complex values for certain values of the argument:

```
>> log(-1)

ans =
  0 + 3.1416i
```

You can use MATLAB to do computations involving complex numbers by entering numbers in the form **a + b*i**:

```
>> (2 + 3*i)*(4 - i)

ans =
  11.0000 +10.0000i
```

Complex arithmetic is a powerful and valuable feature. Even if you don't intend to use complex numbers, you should be alert to the possibility of complex-valued answers when evaluating MATLAB expressions.

More on Matrices

In addition to the usual algebraic methods of combining matrices (e.g., matrix multiplication), we can also combine them element-wise. Specifically, if **A** and **B** are the same size, then **A.*B** is the *element-by-element* product of **A** and **B**, i.e., the matrix whose i, j element is the product of the i, j elements of **A** and **B**. Likewise, **A./B** is the element-by-element quotient of **A** and **B**, and **A.^c** is the matrix formed by raising each of the elements of **A** to the power **c**. More generally, if **f** is one of the built-in mathematical functions in MATLAB, or is a user-defined vectorized function, then **f(A)** is the matrix obtained by applying **f** element-by-element to **A**. See what happens when you type **sqrt(A)**, where **A** is the matrix defined at the beginning of the *Matrices* section of Chapter 2.

Recall that **x(3)** is the third element of a vector **x**. Likewise, **A(2,3)** represents the 2,3 element of **A**, i.e., the element in the second row and third column. You can specify submatrices in a similar way. Typing **A(2,[2 4])** yields the second and fourth elements of the second row of **A**. To select the second, third, and fourth elements of this row, type **A(2,2:4)**. The submatrix consisting of the elements in rows 2 and 3 and in columns 2, 3, and 4 is generated by **A(2:3,2:4)**. A colon by itself denotes an entire row or column. For example, **A(:,2)** denotes the second column of **A**, and **A(3,:)** yields the third row of **A**.

MATLAB has several commands that generate special matrices. The commands **zeros(n,m)** and **ones(n,m)** produce $n \times m$ matrices of zeros and ones, respectively. Also, **eye(n)** represents the $n \times n$ identity matrix.

Solving Linear Systems

Suppose that **A** is a non-singular $n \times n$ matrix and **b** is a column vector of length n. Then typing **x = A\b** numerically computes the unique solution to **A*x = b**. Type **help mldivide** for more information.

If either **A** or **b** is symbolic rather than numerical, then **x = A\b** computes the solution to **A*x = b** symbolically. To calculate a symbolic solution when both inputs are numerical, type **x = sym(A)\b**.

Calculating Eigenvalues and Eigenvectors

The eigenvalues of a square matrix **A** are calculated with **eig(A)**. The command **[U, R] = eig(A)** calculates both the eigenvalues and eigenvectors. The eigenvalues are the diagonal elements of the diagonal matrix **R**, and the columns of **U** are the eigenvectors. Here is an example illustrating the use of **eig**:

```
>> A = [3 -2 0; 2 -2 0; 0 1 1];
>> eig(A)

ans =
    1
   -1
    2

>> [U, R] = eig(A)

U =
         0   -0.4082    -0.8165
         0   -0.8165    -0.4082
    1.0000    0.4082    -0.4082

R =
    1    0    0
    0   -1    0
    0    0    2
```

The eigenvector in the first column of **U** corresponds to the eigenvalue in the first column of **R**, and so on. These are numerical values for the eigenpairs. Note that MATLAB has normalized **U** so that each of its columns is a unit vector. To get symbolically calculated eigenpairs, type **[U, R] = eig(sym(A))**.

Doing Calculus with MATLAB

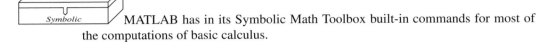 MATLAB has in its Symbolic Math Toolbox built-in commands for most of the computations of basic calculus.

Differentiation

You can use **diff** to differentiate symbolic expressions, and also to approximate the derivative of a function given numerically (say by an M-file):

```
>> syms x, diff(x^3)

ans =
3*x^2
```

Here MATLAB has figured out that the variable is **x**. (See *Default Variables* at the end of the chapter.) Alternatively:

```
>> f = @(x) x^3; diff(f(x))

ans =
3*x^2
```

You can obtain second derivatives with **diff(f(x), 2)** and nth derivatives with **diff(f(x), n)**. The command **diff** can also compute partial derivatives of expressions involving several variables, as in **diff(x^2*y, y)**, but to compute multiple partials with respect to mixed variables you must use **diff** repeatedly, as in **diff(diff(sin(x*y/z), x), y))**. (Remember to declare **y** and **z** to be symbolic.)

Differential equations can be solved with the Symbolic Math Toolbox using the command **dsolve**. Two kinds of syntax are allowed by this command. You can give the input as a string, representing differentiation by the letter **D**, or else you can represent differentiation by **diff** and write the "equals sign" in the equation as a *double* equals sign **==**. To use this second syntax, you need to define the dependent variable as a symbolic function. For example, to use the symbolic ODE solver on the differential equation $xy' + 1 = y$, you can enter either

```
>> dsolve('x*Dy + 1 = y', 'x')

ans =
C5*x + 1
```

or else:

```
>> syms y(x); dsolve(x*diff(y) + 1 == y, x)

ans =
C5*x + 1
```

Integration

MATLAB can compute definite and indefinite integrals. Here is an indefinite integral:

```
>> syms x; int(x^2)
```

```
ans =
x^3/3
```

Note that MATLAB does not include a constant of integration; the output is a single antiderivative of the integrand. Now here is a definite integral:

```
>> int(asin(x), 0, 1)
```

```
ans =
pi/2 - 1
```

You are undoubtedly aware that not every function that appears in calculus can be symbolically integrated, and so numerical integration is sometimes necessary. You can accomplish this by applying **double** to a symbolic integral. Alternatively, MATLAB has several commands for numerical integration of a function $f(x)$, including **quad** and **quadl**, but the best for general purposes is **integral**. You can improve the accuracy by changing the options **'AbsTol'** and **'RelTol'**, the absolute and relative error tolerances, or as below by using **'Waypoints'** to force the integration to go through certain specific points:

```
>> hardintegral = int(log(1 + x^2)*exp(-x^2), x, 0, 1)
```

```
hardintegral =
int(exp(-x^2)*log(x^2 + 1)), x == 0..1)
```

```
>> format long, double(hardintegral)
```

```
ans =
   0.153886113452273
```

```
>> integral(@(x) log(1 + x.^2).*exp(-x.^2), 0, 1)
```

```
ans =
   0.153886113452273
```

```
>> integral(@(x) abs(x), -1, 2)
```

```
ans =
   2.500000418100217
```

```
>> integral(@(x) abs(x), -1, 2, 'Waypoints', 0)
```

```
ans =
   2.500000000000000
```

⇨   **The commands quad and quadl will not accept Inf or -Inf as a limit of integration, but int and integral will. So you can use MATLAB to compute improper integrals, either symbolically or numerically.**

MATLAB can also do multiple integrals. The following command computes the double integral

$$\int_0^\pi \int_0^{\sin x} (x^2 + y^2)\, dy\, dx :$$

```
>> syms x y; int(int(x^2 + y^2, y, 0, sin(x)), 0, pi)
```

```
ans =
pi^2 - 32/9
```

Note that MATLAB presumes that the variable of integration in **int** is **x** unless you prescribe otherwise. Note also that the order of integration is as in calculus, from the "inside out." Finally, we observe that there is a numerical double-integral command **integral2**, whose properties and use we encourage you to discover from the online help.

Limits

You can use **limit** to compute ordinary limits as well as right- and left-handed limits and limits at infinity. For example, here is $\lim_{x \to 0} \sin(x)/x$:

```
>> syms x; limit(sin(x)/x, x, 0)

ans =
1
```

To compute one-sided limits, use the **'right'** and **'left'** options. For example:

```
>> limit(abs(x)/x, x, 0, 'left')

ans =
-1
```

Limits at infinity can be computed using the symbol **Inf**:

```
>> limit((x^4 + x^2 - 3)/(3*x^4 - log(x)), x, Inf)

ans =
1/3
```

Sums and Products

Finite numerical sums and products can be computed easily using the vector capabilities of MATLAB and the commands **sum** and **prod**. For example:

```
>> X = 1:7;
>> sum(X)

ans =
   28

>> prod(X)

ans =
   5040
```

You can do finite and infinite symbolic sums using the command **symsum**. To illustrate, here is the well-known infinite sum

$$\sum_{n=1}^{\infty} \frac{1}{n^2} :$$

```
>> symsum(1/n^2, 1, Inf)
```

```
ans =
pi^2/6
```

Another familiar example is the sum of the infinite geometric series. Since the series is only convergent when the ratio parameter a satisfies $-1 < a < 1$, we can build that assumption in with the command **assume**:

```
>> syms a k; assume(-1 < a < 1); symsum(a^k, 0, Inf)
```

```
ans =
-1/(a-1)
```

Note that if you **clear** the assumption on a, you get a more complicated answer:

```
>> syms a clear; symsum(a^k, 0, Inf)
```

```
ans =
piecewise([1 <= a, Inf], [abs(a) < 1, -1/(a - 1)])
```

This indicates that the sum is ∞ for $a \geq 1$ and has the value we obtained before in the interval of convergence $|a| < 1$.

Taylor Series

You can use **taylor** to generate Taylor polynomial expansions of a specified degree at a specified point (assumed to be 0 if not specified). For example, to generate the Taylor polynomial up to degree 9 at $x = 0$ of the function $\sin x$, we enter:

```
>> syms x; taylor(sin(x), 'Order', 10)
```

```
ans =
x^9/362880 - x^7/5040 + x^5/120 - x^3/6 + x
```

Note that the command **taylor(sin(x), 'Order', 11)** would give the same result, since there is no term of degree 10 in the Taylor expansion of $\sin x$. You can also compute a Taylor polynomial at a point other than the origin. For example:

```
>> taylor(sqrt(x), 'ExpansionPoint', 4, 'Order', 3)
```

```
ans =
x/4 - (x - 4)^2/64 + 1
```

computes a Taylor polynomial of $\sqrt{x}$ up to terms of order 2 centered at the point $x = 4$.

The command **taylor** can also compute Taylor expansions at infinity:

```
>> taylor(exp(1/x^2), x, Inf)
```

```
ans =
1/x^2 + 1/(2*x^4) + 1
```

Default Variables

You can use any letters to denote variables in functions – either MATLAB's or the ones you define. For example, there is nothing special about the use of **t** in the following, any letter will do as well:

```
>> syms t; diff(sin(t^2))
ans =
2*t*cos(t^2)
```

However, if there are multiple variables in an expression and you employ a MATLAB command that does not make explicit reference to one of them, then either you must make the reference explicit or MATLAB will use a built-in hierarchy to decide which variable is the "one in play." For example: **solve('x + y = 3')** solves for **x**, not **y**. If you want to solve for **y** in this example, you need to enter **solve('x + y = 3', 'y')**. MATLAB's default variable for **solve** is **x**. If there is no **x** in the equation(s) to solve, MATLAB looks for the nearest letter to **x** (where **y** takes precedence over **w**, but **w** takes precedence over **z**, etc.). Similarly for **diff**, **int**, and many other symbolic commands. Thus **syms w z; diff w*z** yields **z** as an answer. On occasion MATLAB assigns a different primary default variable – for example, the default independent variable for MATLAB's symbolic ODE solver **dsolve** is **t**. This is mentioned clearly in the online help. If you have doubt about the default variables for any MATLAB command, you should check the online help.

✓   Calculations with the Symbolic Math Toolbox are in fact sent by MATLAB to another program called MuPAD for processing. The MuPAD kernel or symbolic engine, as it is called, performs the symbolic calculation and sends the result back to MATLAB. On rare occasions, you may want to access the MuPAD kernel directly. This can be done with the **evalin** command (or the **feval** command, which is similar) as in the following example:

```
>> evalin(symengine, 'nextprime(10^20)')
ans =
100000000000000000039
```

Here we have invoked the MuPAD command **nextprime**, which has no equivalent in MATLAB, to find the next prime after 10^{20}. The reference to **symengine** tells MATLAB that this command is to be evaluated in the symbolic engine. More details about the MuPAD kernel will be provided in Chapter 8.

Chapter 5

MATLAB Graphics

In this chapter, we describe more of MATLAB's graphics commands and the most common ways of manipulating and customizing graphics. For an overview of commands, type **help graphics** (for general graphics commands), **help graph2d** (for two-dimensional graphics commands), **help graph3d** (for three-dimensional graphics commands), or **help specgraph** (for specialized graphing commands).

We have already discussed the commands **plot** and **ezplot** in Chapter 2. We will begin this chapter by discussing more uses of these commands, as well as some of the other most commonly used plotting commands. Then we will discuss methods for customizing and manipulating graphics. Finally, we will introduce some commands and techniques for creating and modifying images and sounds.

✓ For most types of graphs that we describe below, there is a command like **plot** that draws the graph from numerical data, and a command like **ezplot** that graphs functions specified by string or symbolic input. The latter commands may be easier to use at first, but are more limited in their capabilities and less amenable to customization. Thus, we emphasize the commands that plot data, which are likely to be more useful to you in the long run. In many cases, there are also shortcuts to the plotting commands via the PLOTS tab at the top of MATLAB Desktop. Later in the chapter, we will describe how to use these shortcuts.

Two-Dimensional Plots

Often we want to draw a curve in the x-y plane, but with y not given explicitly as a function of x. There are two main techniques for plotting such curves: parametric plotting and contour or implicit plotting.

Parametric Plots

Sometimes x and y are both given as functions of some parameter. For example, the circle of radius 1 centered at $(0,0)$ can be expressed in *parametric* form as $x = \cos(2\pi t)$, $y = \sin(2\pi t)$, where t runs from 0 to 1. Though y is not expressed as a function of x, you can easily graph this curve with **plot**, as follows (see Figure 5.1):

```
>> T = 0:0.01:1;
>> plot(cos(2*pi*T), sin(2*pi*T)); axis square
```

If we had used an increment of only 0.1 in the **T** vector, the result would have been a polygon with clearly visible corners. When your graph has corners that shouldn't be

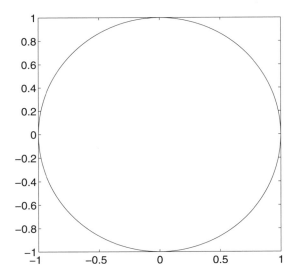

Figure 5.1. The Unit Circle $x^2 + y^2 = 1$.

present, you should repeat the process with a smaller increment until you get a graph that looks smooth. Here also **axis square** forces the same scale on both axes; without it the circle would look like an ellipse.

Parametric plotting is also possible with **ezplot**. You can obtain almost the same graph as in Figure 5.1 with the command:

```
>> ezplot('cos(t)', 'sin(t)', [0 2*pi]); axis square
```

Notice that we used a semicolon after the **ezplot** command, but it did not prevent the graph from appearing. In general, the semicolon suppresses only text output.

Contour Plots and Implicit Plots

A *contour plot* of a function of two variables is a plot of the *level curves* of the function, i.e., sets of points in the x-y plane where the function assumes a constant value. For example, the level curves of $x^2 + y^2$ are circles centered at the origin, and the *levels* are the squares of the radii of the circles. Contour plots are produced in MATLAB with **meshgrid** and **contour**. The command **meshgrid** produces a grid of points in a rectangular region, with a specified spacing. This grid is used by **contour** to produce a contour plot in the specified region.

You can make a contour plot of $x^2 + y^2$ as follows (see Figure 5.2):

```
>> [X Y] = meshgrid(-3:0.1:3, -3:0.1:3);
>> contour(X, Y, X.^2 + Y.^2); axis square
```

You can specify particular level sets by including an additional vector argument to **contour**. For example, to plot the circles of radius 1, $\sqrt{2}$, and $\sqrt{3}$, type:

```
>> contour(X, Y, X.^2 + Y.^2, [1 2 3])
```

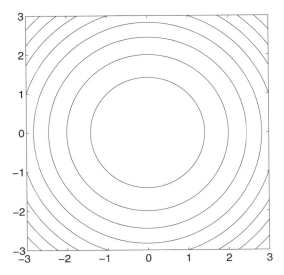

Figure 5.2. Contour Plot of $x^2 + y^2$.

The vector argument must contain at least two elements, so, if you want to plot a single level set, you must specify the same level twice. This is quite useful for implicit plotting of a curve given by an equation in x and y. For example, to plot the circle of radius 1 about the origin, type:

```
>> contour(X, Y, X.^2 + Y.^2, [1 1])
```

Or to plot the lemniscate $x^2 - y^2 = (x^2 + y^2)^2$, rewrite the equation as

$$(x^2 + y^2)^2 - x^2 + y^2 = 0$$

and type (see Figure 5.3):

```
>> [X Y] = meshgrid(-1.1:0.01:1.1, -1.1:0.01:1.1);
>> contour(X, Y, (X.^2 + Y.^2).^2 - X.^2 + Y.^2, [0 0])
>> title('The lemniscate x^2-y^2=(x^2+y^2)^2'); axis square
```

☞ *In this case, we used ^ to produce exponents in the title. You can also use _ for subscripts, and to produce a Greek letter, precede its name with a backslash – for example,* **\theta**. *Type* **doc title** *and look under "Examples" for other tricks with titles; these features also apply to labeling commands such as* **xlabel** *and* **ylabel**. *For more on annotating graphs, see the section* Customizing Graphics *later in this chapter.*

You can also do contour plotting with the command **ezcontour**, and implicit plotting of a curve $f(x, y) = 0$ with **ezplot**. In particular, you can obtain essentially the same picture as Figure 5.2 with the command:

```
>> ezcontour('x^2 + y^2', [-3, 3], [-3, 3]); axis square
```

and almost the same picture as Figure 5.3 with the command:

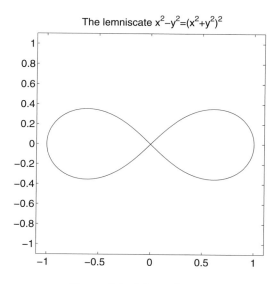

The lemniscate $x^2-y^2=(x^2+y^2)^2$

Figure 5.3. A Lemniscate.

```
>> ezplot('(x^2 + y^2)^2 - x^2 + y^2', ...
[-1.1, 1.1], [-1.1, 1.1]); axis square
```

Field Plots

The MATLAB routine **quiver** is used to plot vector fields or arrays of arrows. The arrows can either be located at equally spaced points in the plane (if x- and y-coordinates are not given explicitly), or they can be placed at specified locations. Sometimes some fiddling is required to scale the arrows so that they don't come out looking too big or too small. For this purpose, **quiver** takes an optional scale-factor argument. The following code, for example, plots a vector field with a "saddle point," corresponding to a combination of an attractive force pointing toward the x-axis and a repulsive force pointing away from the y-axis (see Figure 5.4).

```
>> [x, y] = meshgrid(-1.1:0.2:1.1, -1.1:0.2:1.1);
>> quiver(x, -y); axis equal off
```

Three-Dimensional Plots

MATLAB has several routines for producing three-dimensional plots.

Curves in Three-Dimensional Space

For plotting curves in 3-space, the basic command is **plot3**. It works like **plot**, except that it takes three vectors instead of two: one for the x-coordinates, one for the

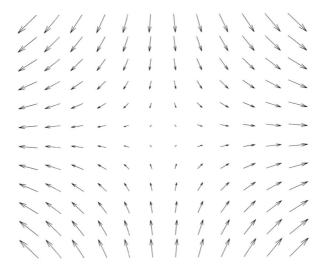

Figure 5.4. A Vector-Field Plot of $(x, -y)$.

y-coordinates, and one for the z-coordinates. For example, we can plot a helix with:

```
>> T = -2:0.01:2;
>> plot3(cos(2*pi*T), sin(2*pi*T), T)
```

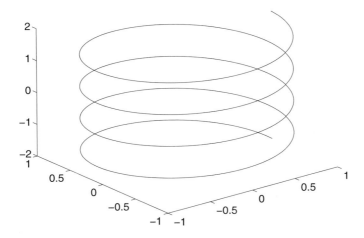

Figure 5.5. The Helix $x = \cos(2\pi z)$, $y = \sin(2\pi z)$.

There is also a three-dimensional analog to **ezplot** called **ezplot3**; you can instead plot the helix in Figure 5.5 with:

```
>> ezplot3('cos(2*pi*t)', 'sin(2*pi*t)', 't', [-2, 2])
```

Surfaces in Three-Dimensional Space

There are two basic commands for plotting surfaces in 3-space: **mesh** and **surf**. The former produces a transparent "mesh" surface, the latter an opaque shaded one. There are two different ways of using each command, one for plotting surfaces in which the z-coordinate is given as a function of x and y, and one for *parametric surfaces* in which x, y, and z are all given as functions of two other parameters. Let us illustrate the former with **mesh** and the latter with **surf**.

To plot $z = f(x, y)$, we begin with a **meshgrid** command as in the case of **contour**. For example, the "saddle surface" $z = x^2 - y^2$ can be plotted with:

```
>> [X,Y] = meshgrid(-2:0.1:2, -2:0.1:2);
>> Z = X.^2 - Y.^2; mesh(X, Y, Z)
```

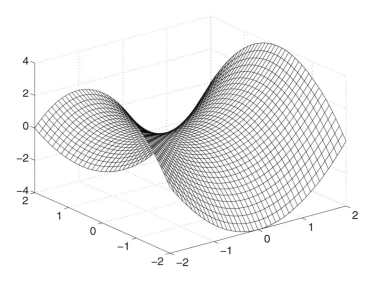

Figure 5.6. The Surface $z = x^2 - y^2$.

The resulting graph (Figure 5.6) looks better on the computer screen, since MATLAB shades the surface with a color scheme depending on the z-coordinate. We could have produced an opaque surface instead by replacing **mesh** with **surf**.

✓   Another option is to click on the three arrays **X**, **Y**, and **Z** in the Workspace Browser, then to click on **surf**, **mesh**, or **surfl** (a variant with different lighting) in the PLOTS tab at the top of the Desktop. The order of selection of the variables matters, and determines which one corresponds to which coordinate axis of the plot.

There are also shortcut commands **ezmesh** and **ezsurf**; you can obtain a result very similar to Figure 5.6 with:

```
>> ezmesh('x^2 - y^2', [-2, 2, -2, 2])
```

If you want to plot a surface that cannot be represented by an equation of the form $z = f(x, y)$, for example the sphere $x^2 + y^2 + z^2 = 1$, then it is better to parameterize the surface using a suitable coordinate system, in this case cylindrical or spherical coordinates. For example, we can take as parameters the vertical coordinate z and the polar coordinate θ in the x-y plane. If r denotes the distance to the z-axis, then the equation of the sphere becomes $r^2 + z^2 = 1$, or $r = \sqrt{1 - z^2}$, and so $x = \sqrt{1 - z^2} \cos \theta$, $y = \sqrt{1 - z^2} \sin \theta$. Thus we can produce our plot of the sphere (Figure 5.7) with:

```
>> [Z, Theta] = meshgrid(-1:0.1:1, (0:0.1:2)*pi);
>> X = sqrt(1 - Z.^2).*cos(Theta);
>> Y = sqrt(1 - Z.^2).*sin(Theta);
>> surf(X, Y, Z); axis square
```

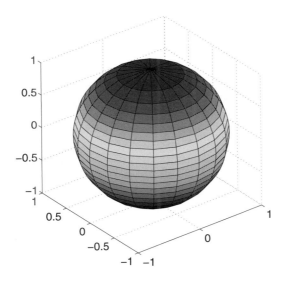

Figure 5.7. The Unit Sphere $x^2 + y^2 + z^2 = 1$.

You can also plot a surface parametrically with **ezmesh** or **ezsurf**; you can obtain a result very similar to Figure 5.7 with:

```
>> ezsurf('sqrt(1-z^2)*cos(t)', ...
'sqrt(1-z^2)*sin(t)', 'z', [0, 2*pi, -1, 1]); axis equal
```

Notice that we had to specify the **t** range first because it is alphabetically before **z**, even though **z** occurs before **t** in the strings we entered. Rather than remember such a rule, you can enter functions of more than one variable as anonymous functions, for example **@(z,t) sqrt(1-z.^2).*cos(t)**; then, since you used **z** as the first input variable, you would specify its range first.

Figure Windows

So far we have discussed only graphics commands that produce or modify a single plot. But MATLAB is also capable of opening multiple figure windows or combining several plots in one figure window. Here we discuss some basic methods for managing and manipulating figure windows.

Multiple Figure Windows

When you execute the first plotting command in a given MATLAB session, the graph appears in a new window labeled "Figure 1". Subsequent graphics commands either modify or replace the graph in this window. You have seen that **hold on** directs that new plotting commands should add to, rather than replace, the current graph. If instead you want to create a new graph in a separate window while keeping the old graph on your screen, type **figure** to open a new window, labeled "Figure 2". Alternatively, you can select **New Figure** from the **File** menu in either your main MATLAB window or the first figure window. Subsequent graphics commands will affect only this window, until you change your *current figure* again with another **figure** command (e.g., type **figure(1)** to switch the current figure back to "Figure 1"), or by bringing another figure to the foreground with the mouse. When multiple figure windows are open, you can find out which is the current one by typing **gcf**, short for "get current figure." Finally, you can delete figure windows in the usual manner with the mouse, or with the command **close**; see the online help for details.

The Figure Toolbar

Each figure window has a toolbar underneath its menu bar with shortcuts for several menu items, including, on the left, icons for opening, saving, and printing figures. Near the middle, there are several icons that correspond to items in the **Tools** menu.

The two icons with plus and minus signs control zooming in and out. Click on the icon with a plus sign, and then click on a point in your graph to zoom in near that point. You can click and zoom multiple times; each zoom changes the scale on both axes by a factor of roughly 2. However, don't click too fast, because double-clicking resets the graph to its original state. Clicking on the icon with the minus sign allows you to zoom out gradually. To the right of the "zoom" and "zoom out" icons is an icon in the shape of a hand that allows you to click and drag the graph to pan both horizontally and vertically within the current axes. More zooming options are available by right-clicking in the figure window while zooming, by using the **Options** submenu of the **Tools** menu, or by using the command **zoom** (see its online help).

Clicking the next icon to the right with the circular arrow allows you to rotate three-dimensional (3D) graphics. For more 3D effects, select the Camera Toolbar from the **View** menu. You can also change the viewpoint with the command **view**. In particular, the command **view(2)** projects a figure into the x-y plane (by looking down on it from the positive z-axis), and the command **view(3)** views it from the default direction in 3-space, which is in the direction looking toward the origin from a point far out on the ray $x = -0.5272t$, $y = -0.6871t$, $z = 0.5t$, $t > 0$.

✓ In MATLAB, any two-dimensional plot can be "viewed in 3D," and any three-dimensional plot can be projected into the plane. Thus Figure 5.5 above (the helix), if followed by the command **view(2)**, produces a circle.

The next icon to the right of the rotate icon enables the Data Cursor, which allows you to display the coordinates of a point on a curve by clicking on or near the curve. Right-clicking in the figure window gives several options; changing Selection Style to Mouse Position will allow you to click on an arbitrary point on the curve rather than just a data point. (Remember that the curves plotted by MATLAB are piecewise-linear curves connecting a finite number of data points.) This can be useful especially after zooming, because data points may then be spaced far apart. Another way to get coordinates of a point is by typing **ginput(1)** in the Command Window; this allows you to click on any point in the figure window, not just on a curve. While this gives you more flexibility, if you want the precise coordinates of a point on a curve, it is best to use the Data Cursor, because it will always highlight a point on the curve even if you don't click exactly on the curve.

The rightmost icon on the Figure Toolbar opens three windows surrounding the figure; these are collectively known as Plot Tools, and you can also open them with the command **plottools**. You can also control the display of these three windows – the Figure Palette, the Plot Browser, and the Property Editor – individually from the **View** menu. These windows enable various means of editing figures. Many of these capabilities are also available in the **Insert** and **Tools** menus, and from the Plot Edit Toolbar in the **View** menu. We will briefly discuss some editing options in the *Customizing Graphics* section below, but there are many more possibilities and we encourage you to experiment with these tools.

Combining Plots in One Window

The command **subplot** divides the figure window into an array of smaller plots. The first two arguments give the dimensions of the array of subplots, and the last argument gives the number of the subplot (counting from left to right across the first row, then from left to right across the next row, and so on) in which to put the output of the next graphing command. The following example, whose output appears in Figure 5.8, produces a 2×2 array of plots of the first four Bessel functions J_n, $0 \le n \le 3$:

```
>> X = 0:0.05:40;
>> for n = 1:4
     subplot(2,2,n)
     plot(X, besselj(n - 1, X))
   end
```

✓ In MATLAB, you can also create subplots using the Figure Palette, which you can enable from the **View** menu or as part of the Plot Tools described above.

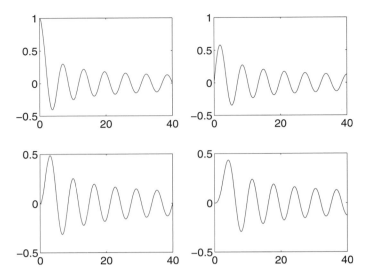

Figure 5.8. Bessel Functions $J_0(x)$ (upper left), $J_1(x)$ (upper right), $J_2(x)$ (lower left), and $J_3(x)$ (lower right).

✰ Customizing Graphics

☞   *This is a more advanced topic; if you wish, you can skip it on a first reading.*

So far in this chapter, we have discussed some commonly used MATLAB routines for generating and manipulating plots. But often, to get a more precise result, you need to customize or edit the graphics these commands produce. In order to do this, you must understand a few basic principles concerning the way MATLAB stores and displays graphics. For most purposes, the discussion here will be sufficient. But if you need more information, you may want to consult the topic "Graphics Objects" in the Graphics section of the MATLAB documentation in the in the Help Browser.

Once you have created a figure, there are two basic ways to manipulate it. The current figure can be modified by typing MATLAB commands in the Command Window, such as the commands **title** and **axis square** that we have already encountered. Or you can modify the figure with the mouse, using the menus and icons in the figure window itself. Almost all of the text commands have counterparts that can be executed directly in the figure window. So why bother learning both techniques? The reason is that editing in the figure window is often more convenient, especially if you wish to "experiment" with multiple changes, while editing a figure with MATLAB commands in an M-file makes your customizations reproducible. So the true MATLAB expert uses both techniques. While the text commands generally remain the same from one version of MATLAB to the next, the figure-window menus tend to change with revisions of MATLAB, and may seem slightly different in the version you are using than in our description here, which is based on MATLAB 2013b.

To modify objects in the figure window with the mouse, editing must be enabled in that window. You can enable or disable editing by selecting **Tools:Edit Plot** or by clicking the arrow icon to the right of the print icon. When editing is enabled, this arrow icon is highlighted, and there is a check mark next to **Edit Plot** in the **Tools** menu. In several places below, we will tell you to click on an object in the figure window in order to edit it. When you click on the object, it should be highlighted with small black squares. If this doesn't happen, then you need to enable editing.

Annotation

In order to insert labels or text into a plot, you can use the commands **text**, **xlabel**, **ylabel**, **zlabel**, and **legend**, in addition to **title**. As the names suggest, **xlabel**, **ylabel**, and **zlabel** add text next to the coordinate axes, **legend** puts a "legend" on the plot, and **text** adds text at a specific point. These commands take various optional arguments that can be used to change the font family and font size of the text. As an example, let's illustrate how to modify our plot of the lemniscate (Figure 5.3) by adding and modifying text:

```
>> title('The lemniscate x^2-y^2=(x^2+y^2)^2', 'FontSize', ...
20, 'FontName', 'Helvetica', 'FontWeight', 'bold')
>> text(0, 0, ' \leftarrow a node, also an inflection', ...
'FontSize', 12)
>> text(0.2, -0.1, 'point for each branch', 'FontSize', 12)
>> xlabel x, ylabel y
```

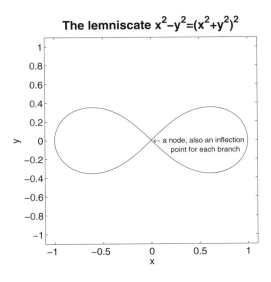

Figure 5.9. The Lemniscate from Figure 5.3 with Annotation and a Larger Title.

Notice that many symbols (such as the arrow pointing to the left in Figure 5.9) can be inserted into a text string by calling them with names starting with the backslash symbol \. (If you've used the scientific typesetting program TEX, you'll recognize the convention here.) In most cases, the names are self-explanatory. For example, you get

a Greek π by typing **\pi**, a summation sign $\sum$ by typing either **\Sigma** (for a capital 'sigma') or **\sum**, and arrows pointing in various directions with **\leftarrow**,
\uparrow, and so on. For more details and a complete list of available symbols, see
the listing for "Text Properties" in the Help Browser (you can find this listing from the
"Search" tab in the Help Browser, or type **doc text** and click on "Text Properties"
at the bottom of its page).

Change of Plot Style

Another important way to change the style of graphics is to modify the color or line
style in a plot or to change the tick marks and labeling on the axes. Within a **plot**
command, you can change the color of a graph, or plot with a dashed or dotted line, or
mark the plotted points with special symbols simply by adding a string third argument
for every x-y pair. Symbols for colors are **'y'** for yellow, **'m'** for magenta, **'c'** for
cyan, **'r'** for red, **'g'** for green, **'b'** for blue, **'w'** for white, and **'k'** for black.
Symbols for point markers include **'o'** for a circle, **'x'** for a cross, **'+'** for a plus
sign, and **'*'** for a star. Symbols for line styles include **'-'** for a solid line, **':'**
for a dotted line, **'--'** for a dashed line. If a point style is given but no line style,
then the points are plotted but no curve is drawn connecting them. (The same methods
work with **plot3** in place of **plot**.) For example, you can produce a solid red sine
curve together with a dotted blue cosine curve, marking all the local maximum points
on each curve with a distinctive symbol of the same color as the plot, as follows:

```
>> X = (-2:0.02:2)*pi; Y1 = sin(X); Y2 = cos(X);
>> plot(X, Y1, 'r-', X, Y2, 'b:'); hold on
>> X1 = [-3*pi/2 pi/2]; Y3 = [1 1]; plot(X1, Y3, 'r*')
>> X2 = [-2*pi 0 2*pi]; Y4 = [1 1 1]; plot(X2, Y4, 'b+')
>> axis([-7 7 -1.1 1.1])
```

 Typing **help plot** gives a useful table of possible line types, plot symbols,
and colors.

Here you may want the tick marks on the x-axis located at multiples of π. This
can be done with the command **set**, which is used to change various properties of
graphics. To apply it to the axes, it has to be combined with the command **gca**, which
stands for "get current axes." The code

```
>> set(gca, 'XTick', (-2:2)*pi, 'XTickLabel', ...
'-2pi|-pi|0|pi|2pi', 'FontSize', 16)
```

in combination with the code above gets the current axes, sets the ticks on the x-axis to
go from -2π to 2π in multiples of π, and then labels these ticks symbolically (rather
than in decimal notation, which is ugly here). It also increases the size of the labels to
a 16-point font. The result is shown in Figure 5.10.

Incidentally, you might wonder how to label the ticks as -2π, $-\pi$, etc., instead of
-2pi, **-pi**, and so on. This is trickier, but you can do it by typing

```
>> set(gca, 'FontName', 'Symbol')
>> set(gca, 'XTickLabel', '-2p|-p|0|p|2p')
```

since in the Symbol font, π occupies the slot held by **p** in text fonts.

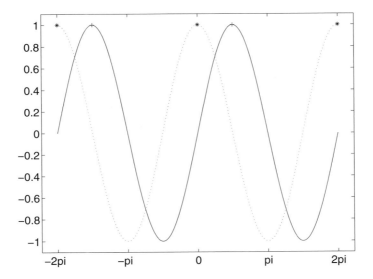

Figure 5.10. Two Periods of $\sin x$ and $\cos x$.

✓ You can again use the Property Editor to make the same types of stylistic changes. Click on a curve and go to the Property Editor to change its style, color, width, etc. In a 3D plot, you can click on a surface and see options for changing its coloring and other properties. To change the tick marks, labeling, etc., click on the axes or a blank area inside them to focus the Property Editor on the axes. Within a figure window, you can also go to **Edit:Current Object Properties...** to modify its properties, or select **Edit:Axes Properties...** to change the font for the tick labels.

Full-Fledged Customization

What about changes to other aspects of a plot? The commands **get** and **set** can be used to obtain a complete list of the properties of the objects in a figure window and then to modify them. These objects and properties are arranged in a hierarchical structure, with each object identified by a floating-point number called a *handle*. If you type **get(gcf)**, you will "get" a (rather long) list of properties of the current figure (whose handle is returned by the function **gcf**). Some of these might read:

```
Color = [0.8 0.8 0.8]
CurrentAxes = [151.001]
Children = [151.001]
```

Here **Color** gives the background color of the plot in red-green-blue (RGB) coordinates, where [0 0 0] is black and [1 1 1] is white; [0.8 0.8 0.8] is light gray. Notice that **CurrentAxes** and **Children** in this example have the same value, the one-element vector containing the funny-looking number 151.001. This number is the handle of the current axes, which would also be returned by the command **gca** ("get current axes"). The fact that this handle also shows up under **Children** indicates that the axes are "children" of the figure, i.e., they lie one level down in the hier-archical structure. Typing **get(gca)** would then give you a list of axis properties, including the handles of further **Children** such as **Line** objects, within which you would find the **XData** and **YData** encoding the actual plot.

⇨ **In the example above, 151.001 is not the exact value of the axes handle, just its first few decimal places. So, typing get(151.001) would yield an error message. To retrieve the exact value of Children in the example above, type get(gcf, 'Children'). Sometimes a figure will have multiple children, in which case this command will return a vector of handles. In general, you should not refer to a handle in an M-file by its numerical value, but instead you should get the handle number via one of the commands get, gcf, gca, or findobj, since handle numbers might change the next time you run the M-file.**

Once you have located the properties you're interested in, they can be changed with **set**. For example,

```
>> set(gcf, 'Color', [1 0 0])
```

changes the background color of the border of the figure window to red, and

```
>> set(gca, 'Color', [1 1 0])
```

changes the background color of the plot itself (a child of the figure window) to yellow (which in the RGB scheme is half red, half green).

This "one at a time" method for locating and modifying figure properties can be speeded up using the command **findobj** to locate the handles of all the descendents (the main figure window, its children, children of children, etc.) of the current figure. You can also limit the search to handles containing elements of a specific type. For example, **findobj('type', 'line')** hunts for all handles of objects containing a Line element. Once you have located these, you can use **set** to change the **LineStyle** from solid to dashed, etc. In addition, the low-level graphics commands **line**, **rectangle**, **fill**, **surface**, and **image** can be used to create new graphics elements within a figure window.

✓ You can also see and modify a full list of properties for the figure, axes, or another object using the Property Editor. Click on the object and then click on the "More Properties..." button in the Property Editor. To select the figure itself, click on the border of the figure, outside the axes.

As an example of these techniques, the following code creates a chessboard on a white background, as shown in Figure 5.11:

```
>> white = [1 1 1]; gray = 0.7*white;
>> a = [0 1 1 0]; b = [0 0 1 1]; c = [1 1 1 1];
>> figure; hold on
>> for k = 0:1, for j = 0:2:6
     fill(a'*c + c'*(0:2:6) + k, b'*c + j + k, gray)
end, end
>> plot(8*a', 8*b', 'k')
>> set(gca, 'XTickLabel', [], 'YTickLabel', [])
>> set(gcf, 'Color', white); axis square
```

Figure 5.11. A Chessboard.

Here **white** and **gray** are the RGB codings for white and gray. The double **for** loop draws the 32 dark squares on the chessboard, using **fill**, with **j** indexing the dark squares in a single vertical column, with **k = 0** giving the odd-numbered rows, and with **k = 1** giving the even-numbered rows. Notice that **fill** here takes three arguments: a matrix, each of whose columns gives the x-coordinates of the vertices of a polygon to be filled (in this case, a square), a second matrix whose corresponding columns give the y-coordinates of the vertices, and a color. We've constructed the matrices with four columns, one for each of the solid squares in a single horizontal row. The **plot** command draws the solid black line around the outside of the board. Finally, the first **set** command removes the printed labels on the axes, and the second **set** command resets the background color to white.

☆ Images, Animations, and Sound

MATLAB is also able to create and manipulate full-color images, animations, and sound files. In addition to the command-line methods described below, you can open a media file in a format that MATLAB supports by double-clicking on it in the Current Folder Browser or by selecting ⬇ **Import Data** in the VARIABLE section of the HOME tab.

Images

MATLAB can read, write, and edit images, such as those created by a digital camera, found on the internet, or created from within MATLAB. An image is simply a two-dimensional array of tiny colored squares called "pixels." The image may be stored in a file in a variety of formats, including PNG ("portable network graphics"), JPEG ("joint photographic experts group"), TIFF ("tagged image file format"), and GIF ("graphics interchange format"). In MATLAB, a color image with height h pixels and width w pixels is generally stored in one of three ways: as an RGB image, as a grayscale image, or as an indexed image. A grayscale image is the simplest type; it is represented as an $h \times w$ array of pixel intensities. An RGB image is represented as an $h \times w \times 3$ array, so that the color of each pixel is specified by three values: a red intensity, a green intensity, and a blue intensity. An indexed image consists of an $h \times w$ array together with an auxiliary $c \times 3$ array called a "colormap"; each element in the first array represents the index of a row in the colormap, and this row gives the RGB values for the corresponding pixel. (What MATLAB calls RGB format is often called "true color" in graphics programming; an indexed image is often called "pseudocolor.")

The command **imread** will read an image from any of a large variety of image-file formats; see the online help for supported formats. Some formats, such as PNG, can store grayscale, RGB, or indexed images. Other formats can store only one or two types; GIF images are always indexed, whereas JPEG images are never indexed. Most images you find on the World Wide Web will be stored in RGB format unless they are GIF files. To read an RGB image from the file `picture.png` and store it in the array **rgbpic**, type:

```
>> rgbpic = imread('picture.png');
```

You can read an indexed image with **imread** by assigning its output to two variables, one for the image array and one for the colormap. You can then convert these arrays to a single RGB array with **ind2rgb**. For example:

```
>> [indpic, map] = imread('picture.gif');
>> rgbpic = ind2rgb(indpic, map);
```

Converting from RGB to indexed format is harder because generally the number of different colors in the image must be reduced; colormaps often have 256 or fewer colors. The command **rgb2ind** offers several algorithms for making this conversion.

The command **image** displays an image in a figure window. For an RGB image **rgbpic**, simply type:

```
>> image(rgbpic)
>> axis equal tight
```

The second command is not necessary, but ensures that the image is displayed in its intended aspect ratio. For an indexed image **indpic** with colormap **map**, type:

```
>> image(indpic)
>> colormap(map)
```

The command **colormap** changes the colormap of the current axes or, with no input arguments, outputs the current colormap.

☞   *For examples of the use of these commands, see the* Image Compression and Manipulation *application in Chapter 11.*

You can edit an image by changing the values in the image array. Notice that when you display an image, the axes are labeled with the indices of the image array. You can use the zoom feature of the figure window (see the *Figure Windows* section earlier in this chapter) to locate more precisely the indices of a particular feature of the image, or even an individual pixel. Theoretically at least, you can then edit the image in any way you want by changing the numbers in the array. Here we describe how to carry out several common manipulations in a practical manner.

⇨   **We will describe mainly how to edit RGB images, since this format allows you more freedom and you can always convert from indexed format to RGB as described above. However, RGB images are stored in three-dimensional arrays, which may take some time to get used to. Previously in this book, we have discussed only two-dimensional arrays, and some MATLAB commands that manipulate two-dimensional arrays do not work for three-dimensional arrays.**

To reverse an image up-to-down or left-to-right, you simply need to reverse the array with **flipdim**. For example,

```
>> image(flipdim(rgbpic, 2))
```

will display a left-to-right mirror image, while **flipdim(rgbpic, 1)** reverses the array up-to-down. (For indexed images and other two-dimensional arrays, you can use the more mnemonic commands **fliplr** and **flipud** instead.)

To crop an image, select the appropriate subarray. For example, to remove 50 pixels from the top and bottom of the image and 100 pixels from the left and right, type:

```
>> newpic = rgbpic(51:end-50, 101:end-100, :);
```

You can then display the cropped image **newpic** as described above, or save it as described below.

To examine the color of an individual pixel, you can display its RGB values in the Command Window and/or display it in a figure window. For example, to examine the pixel of **rgbpic** in the lower left-hand corner, type:

```
>> rgbpic(end, 1, :)
```

```
ans(:,:,1) =
   240
ans(:,:,2) =
   114
ans(:,:,3) =
   14
```

The output above gives (hypothetical) red, green, and blue values for the pixel. To see this color in the current figure window, type **image(rgbpic(end, 1, :))**. You can then adjust the color as desired; for instance, typing **rgbpic(end, 1, 2) = 180** will increase the green intensity, making the color lighter and more yellow.

Of course, changing a single pixel will not change the appearance of the figure much, but you can also change the color of a whole block of pixels or the entire image in a similar manner. For example, to black out a rectangle within the picture, set all of the values in the corresponding subarray to 0. Thus, **rgbpic(40:60, 90:110, :) = 0** will change to black all the pixels in a 21×21 square centered 50 pixels from the top and 100 pixels from the left of the image.

Changing the numbers in the array as we have just described will not change the image displayed in the figure window until you issue a new **image** command.

In addition to manipulating images that you read into MATLAB, you can create your own images to visualize numerical data. Suppose, for example, that you have an array **temp** that contains temperatures for some geographical region. You can display the temperatures as an indexed image by typing:

```
>> imagesc(temp)
```

The command **imagesc** works like **image**, except that it rescales the values in a two-dimensional array so that the highest number corresponds to the highest numbered color in the current colormap and the lowest number corresponds to the lowest numbered color. With the default colormap in effect, hot regions will be colored red and cold regions will be colored blue, with other colors representing intermediate temperatures.

To display the colormap next to the image, type **colorbar**. Notice that the numbers on this bar correspond to the numbers in your original array, and are independent of the rescaling done by **imagesc**. If you want a different colormap, you can create your own or use one of the other colormaps built into MATLAB; type **doc colormap** for a selection. For example, to transform the temperature map described above into a black-and-white image, where white represents hot, black represents cold, and intermediate temperatures are in shades of gray, type **colormap(gray)**.

Finally, you can save an image in one of the standard formats like PNG with the command **imwrite**. For example, to save **newpic** to the file newpict.png, type:

```
>> imwrite(newpic, 'newpict.png')
```

⇨   **In Chapter 3, we discussed how to save a figure in a format such as PNG using either print or File:Save As... from the figure window menu. This will save the entire figure, including the border and axis labels, into the image file. If you only want to save the image inside the axes, use imwrite.**

Animations

The simplest way to produce an animated picture is with **comet**, which produces a parametric plot of a curve (the way **plot** does), except that you can see the curve being traced out in time. For example,

```
>> T = (0:0.01:2)*pi;
>> figure, axis equal, axis([-1 1 -1 1]), hold on
>> comet(cos(T), sin(T))
```

displays uniform circular motion.

 We used **hold on** here not to save a previous graph, but to preserve the axis properties we had just set. Without **hold on**, MATLAB would revert to its default axes, and the curve would look elliptical rather than circular as it is being traced.

For more complicated animations, you can use **getframe** and **movie**. The command **getframe** captures the active figure window for one frame of the movie, and **movie** plays back the result. For example, the following commands produce a movie of a vibrating string:

```
>> X = 0:0.01:1;
>> for n = 0:50
      plot(X, sin(n*pi/5)*sin(pi*X)), axis([0, 1, -2, 2])
      M(n+1) = getframe;
end
>> figure(gcf); movie(M)
```

The **axis** command here is important, to ensure that each frame of the movie is drawn with the same coordinate axes. (Otherwise, the scale of the axes will be different in each frame, and the resulting movie will be totally misleading.) The semicolon after **getframe** is highly recommended but not essential (you can see what happens if you forget it). Similarly, the **figure** command is not necessary, but it brings the current figure window to the foreground before the movie starts playing. The command **movie** for playing back the animation has optional parameters for specifying the number of times to play the movie (the default is 1) and the number of frames per second (the default is 12).

Once you have created a movie, you can use **movie2avi** to save it as an AVI file, which is a standard format that can be used in other movie-viewing programs, such as Windows Media Player and QuickTime. For example, to save the movie created above to the file string.avi, type **movie2avi(M, 'string.avi')**.

☞   *It is also possible to create animations within MuPAD. These will be discussed at the end of Chapter 8. Details may be found by searching under "Animations" in the MuPAD documentation.*

Sound

You can use **sound** to generate sound on your computer (provided that your computer is suitably equipped). This command takes a vector, regards it as the waveform of a sound, and "plays" it. A "sinusoidal" vector corresponds to a pure tone, and the frequency of the sinusoidal signal determines the pitch. Thus, the following example plays the motto from Beethoven's Fifth Symphony:

```
>> x = (0:0.1:250)*pi; y = zeros(1,200); z = (0:0.1:1000)*pi;
>> w = [sin(x), y, sin(x), y, sin(x), y, sin(z*4/5), y, ...
sin(8/9*x), y, sin(8/9*x), y, sin(8/9*x), y, sin(z*3/4)];
>> sound(w)
```

Notice that the zero vector **y** in this example creates a very short pause between successive notes.

For **sound**, the values in the input vector should be between -1 and 1. Within that range, the amplitude of the vector determines the volume of the sound; to play the sound above at half volume, you can multiply the input vector by 0.5. For a vector with an amplitude greater than 1, you can use **soundsc** to rescale the vector to the range -1 to 1 before playing it.

By default, the sound is played at 8192 samples per second, so the length of the vector, divided by 8192, is the length of the sound in seconds. You can change the sample rate with an optional second argument to **sound**; this will change both the pitch and the duration of the sound you hear.

Finally, you can read and write sound files with the command **audiowrite**. Possible output formats include WAV and MP4. Note that you can only write in Windows to MP4 files in a narrow range of sample rates (44100 or 48000). The little Beethoven snippet above can be written to a file with:

```
>> audiowrite('beethoven.wav',w,8192)
```

Chapter 6

MATLAB Programming

Every time you create an M-file, you are writing a computer program using the MATLAB programming language. You can do quite a lot in MATLAB using no more than the most basic programming techniques that we have already introduced. In particular, we discussed simple loops (using **for**) and a rudimentary approach to debugging in Chapter 3. In this chapter, we will cover some further programming commands and techniques that are useful for attacking more complicated problems with MATLAB. If you are already familiar with another programming language, much of this material will be quite easy for you to pick up!

✓ Many MATLAB commands are themselves M-files, which you can examine using **type** or **edit**; e.g., enter **type isprime** to see the M-file for the command **isprime**. You can learn a lot about MATLAB programming techniques by inspecting the built-in M-files. Incidentally, although you can edit a built-in MATLAB M-file, like **isprime**, you cannot save it as a system file; you can rename it and save it as a file for your personal use.

Branching

For many user-defined functions, you can use a function M-file that executes the same sequence of commands for each input. However, you might want a function to perform a different sequence of commands in different cases, depending on the input. You can accomplish this with a branching command, and, as in many other programming languages, branching in MATLAB is usually done with the command **if**, which we will discuss now. Later we will describe the other main branching command, **switch**.

Branching with `if`

For a simple illustration of branching with **if**, consider the following function M-file absval.m, which computes the absolute value of a real number:

```
function y = absval(x)
if x >= 0
    y = x;
else
    y = -x;
end
```

The first line of this M-file states that the function has a single input **x** and a single output **y**. If the input **x** is non-negative, the **if** statement is determined by MATLAB

77

to be true. Then the command between the **if** and the **else** statements is executed to set **y** equal to **x**, while MATLAB skips the command between the **else** and **end** statements. On the other hand, if **x** is negative, then MATLAB skips to the **else** statement and executes the succeeding command, setting **y** equal to **-x**. As with a **for** loop, the indentation of commands above is optional; it is helpful to the human reader and is done automatically by MATLAB's built-in Editor/Debugger.

✓   Most of the examples in this chapter will give peculiar results if their input is of a different type than intended. The M-file absval.m is designed only for scalar real inputs **x**, not for complex numbers or vectors. If **x** is complex for instance, then **x >= 0** checks only whether the real part of **x** is non-negative, and the output **y** will be complex in either case. MATLAB has a built-in function **abs** that works correctly for vectors of complex numbers.

In general, **if** must be followed on the same line by an expression that MATLAB will test to be true or false; see the section below on *Logical Expressions* for a discussion of available expressions and how they are evaluated. After some intervening commands, there must be (as with **for**) an **end** statement. In between, there may be one or more **elseif** statements (see below) and/or an **else** statement (as above). If the result of the test is true, MATLAB executes all commands between the **if** statement and the first **elseif**, **else**, or **end** statement, then skips all other commands until after the **end** statement. If the result of the test is false, MATLAB skips to the first **elseif**, **else**, or **end** statement and proceeds from there, carrying out a new test in the case of an **elseif** statement. In the example below, we reformulate absval.m so that no commands are necessary if the result of the test is false, eliminating the need for an **else** statement:

```
function y = absval(x)
y = x;
if y < 0
    y = -y;
end
```

The **elseif** statement is useful if there are more than two alternatives and they can be distinguished by a sequence of true/false tests. It is essentially equivalent to an **else** statement followed immediately by a nested **if** statement. In the example below, we use **elseif** in an M-file signum.m, which evaluates the function

$$\mathrm{sgn}(x) = \begin{cases} 1 & x > 0, \\ 0 & x = 0, \\ -1 & x < 0. \end{cases}$$

(Again, MATLAB has a built-in function **sign** that implements this function for more general inputs than we consider here.)

```
function y = signum(x)
if x > 0
    y = 1;
elseif x == 0
    y = 0;
else
```

```
    y = -1;
end
```

If the input **x** is positive here, then the output **y** is set to **1** and all commands from the **elseif** statement to the **end** statement are skipped. (In particular, the test in the **elseif** statement is not performed.) If **x** is not positive, then MATLAB skips to the **elseif** statement and tests to see whether **x** equals **0**. If so, **y** is set to **0**, otherwise **y** is set to **-1**. Notice that MATLAB requires a double equals sign **==** to test for equality; a single equals sign is reserved for the assignment of values to variables.

✓   Like **for** and the other programming commands you will encounter, **if** and its associated commands can be used in the Command Window. Doing so can be useful for practice with these commands, but they are intended mainly for use in M-files. In our discussion of branching, we consider primarily the case of function M-files; branching is less often used in script M-files.

Logical Expressions

In the examples above, we used *relational operators* such as **>=**, **>**, and **==** to form a logical expression, and instructed MATLAB to choose between different commands according to whether the expression is true or false. Type **help relop** to see all of the available relational operators. Some of these operators, like **&** (AND) and **|** (OR) can be used to form logical expressions that are more complex than those that simply compare two numbers. For example, the expression **(x > 0) | (y > 0)** will be true if **x** or **y** (or both) is positive, and false if neither is positive. In this particular example, the parentheses are not necessary, but, generally, compound logical expressions like this are both easier to read and less prone to errors if parentheses are used to avoid ambiguities.

So far in our discussion of branching, we have considered only expressions that can be evaluated as true or false. While such expressions are sufficient for many purposes, you can also follow **if** or **elseif** with any expression that MATLAB can evaluate numerically. In fact, MATLAB makes almost no distinction between logical expressions and ordinary numerical expressions. Consider what happens if you type a logical expression by itself in the Command Window:

```
>> 2 > 3
ans =
    0
```

When evaluating a logical expression, MATLAB assigns it a value of **0** (for FALSE) or **1** (for TRUE). Thus if you type **2 < 3**, the answer is **1**. The relational operators are treated by MATLAB like arithmetic operators, inasmuch as their output is numerical.

✓   MATLAB makes a subtle distinction between the output of relational operators and ordinary numbers. For example, if you type **whos** after the command above, you will see that the class of **ans** is not *double*, but instead *logical array*. We will give an example of how this feature can be used shortly. Type **help logical** for more information.

Here is another example:

```
>> 2 | 3
ans =
    1
```

The OR operator | gives the answer **0** if both operands are zero and **1** otherwise. Thus, while the output of relational operators is always **0** or **1**, any non-zero input to operators such as **&** (AND), | (OR), and **~** (NOT) is regarded by MATLAB to be true, while only **0** is regarded to be false.

If the inputs to a relational operator are vectors or matrices rather than scalars, then, as for arithmetic operations such as **+** and **.***, the operation is done term-by-term and the output is an array of zeros and ones. Here are some examples:

```
>> [2 3] < [3 2]
ans =
    1    0
>> x = -2:2; x >= 0
ans =
    0    0    1    1    1
```

In the second case, **x** is compared term-by-term with the scalar **0**. Type **help relop** for more information.

You can use the fact that the output of a relational operator is a logical array to select the elements of an array that meet a certain condition. For example, the expression **x(x >= 0)** yields a vector consisting of only the non-negative elements of **x** (or more precisely, those with non-zero real part). So, if **x = -2:2** as above:

```
>> x(x >= 0)
ans =
    0    1    2
```

If a logical array is used to choose elements from another array, the two arrays must have the same size. The elements corresponding to each **1** in the logical array are selected while the elements corresponding to each **0** are not. In the example above, the result is the same as if we had typed **x(3:5)**, but in this case **3:5** is an ordinary numerical array specifying the numerical indices of the elements to choose.

Next, we discuss how **if** and **elseif** decide whether an expression is true or false. For an expression that evaluates to a scalar real number, the criterion is the same as described above – namely, a non-zero number is treated as true while **0** is treated as false. However, for complex numbers only the real part is considered – thus, in an **if** or **elseif** statement, any number with non-zero real part is treated as true, while numbers with zero real part are treated as false. Furthermore, if the expression evaluates to a vector or matrix, an **if** or **elseif** statement must still result in a single true-or-false decision. The convention MATLAB uses is that all elements must be true – that is, all elements must have non-zero real part – for an expression to be treated as true. If any element has zero real part, then the expression is treated as false.

You can manipulate the way branching is done with vector input by inverting tests with **~** and using the commands **any** and **all**. For example, the statements **if x == 0; ...; end** will execute a block of commands (represented here by **...**) when all the elements of **x** are zero; if you would like to execute a block of commands

when *any* of the elements of **x** is zero, you could use the form **if x ~= 0; else;**
...; end. Here **~=** is the relational operator for "does not equal," so the test fails
when any element of **x** is zero, and execution skips past the **else** statement. You can
achieve the same effect in a more straightforward manner using **any**, which outputs
true when any element of an array is non-zero: **if any(x == 0); ...; end**
(remember that if any element of **x** is zero, the corresponding element of **x == 0** is
non-zero). Likewise, **all** outputs true when all elements of an array are non-zero.

Branching with Logical Expressions

Here is a series of examples to illustrate some of the features of logical expressions
and branching that we have just described. Suppose that you want to create a function
M-file that computes the following function:

$$f(x) = \begin{cases} \sin(x)/x & x \neq 0, \\ 1 & x = 0. \end{cases}$$

You could construct the M-file as follows:

```
function y = f(x)
if x == 0
    y = 1;
else
    y = sin(x)/x;
end
```

This will work fine if the input **x** is a scalar, but not if **x** is a vector or matrix. For
example, if **x** is a vector, then at least two things can go wrong. First, MATLAB
would detect the error if you attempted to multiply two vectors using * (without the
dot (**.**)). But for division, if you omit the dot, MATLAB will try matrix division or
some kind of least squares approximation. This is because the "slash" (**/**) has more
than one meaning in MATLAB (type **help mrdivide** for more information). Of
course, you could change **/** to **./** in the second definition of **y**, and change the first
definition to make **y** the same size as **x**. But, if **x** has both zero and non-zero elements,
then MATLAB will declare the **if** statement to be false, and use the second definition.
Then some of the entries in the output array **y** will be **NaN**, "not a number," because
0/0 is an indeterminate form.

One way to make this M-file work for vectors and matrices is to use a loop to
evaluate the function element-by-element, with an **if** statement inside the loop:

```
function y = f(x)
y = ones(size(x));
for n = 1:prod(size(x))
    if x(n) ~= 0
        y(n) = sin(x(n))/x(n);
    end
end
```

In the M-file above, we first create the eventual output **y** as an array of ones with the
same size as the input **x**. Here we use **size(x)** to determine the number of rows and
columns of **x**; recall that MATLAB treats a scalar or a vector as an array with one row

and/or one column. Then **prod(size(x))** yields the number of elements in **x**. So, in the **for** statement, **n** varies from **1** to this number. For each element **x(n)**, we check to see whether it is non-zero, and if so, we redefine the corresponding element **y(n)** accordingly. (If **x(n)** equals **0**, there is no need to redefine **y(n)**, since we defined it initially to be **1**.)

✓ We just used an important but subtle feature of MATLAB, namely that each element of a matrix can be referred to with a single index; for example, if **x** is a 3×2 array, then its elements can be enumerated as **x(1),x(2),...,x(6)**. In this way, we avoided using a loop within a loop. Similarly, we could use **length(x(:))** in place of **prod(size(x))** to count the total number of entries in **x**. However, we need to be careful. If we had not predefined **y** to have the same size as **x**, but rather used an **else** statement inside the loop to let **y(n)** be **1** when **x(n)** is **0**, then **y** would have ended up a 1×6 array rather than a 3×2 array. We then could have used the command **y = reshape(y, size(x))** at the end of the M-file to make **y** have the same shape as **x**. However, even if the shape of the output array is not important, it is generally best to predefine an array of the appropriate size before computing it element-by-element in a loop, because the loop will then run faster.

Next, consider the following modification of the M-file above:

```
function y = f(x)
if x ~= 0
    y = sin(x)./x;
    return
end
y = ones(size(x));
for n = 1:prod(size(x))
    if x(n) ~= 0
        y(n) = sin(x(n))/x(n);
    end
end
```

Above the loop we added a block of four lines whose purpose is to make the M-file run faster if all the elements of the input **x** are non-zero. Because MATLAB processes vectors more efficiently than loops, the new M-file runs several times faster if **x** has a large number of elements (all non-zero). Here is how the new block of four lines works. The first **if** statement will be true provided that all the elements of **x** are non-zero. In this case, we define the output **y** using MATLAB's vector operations, which are generally much more efficient than running a loop. Then we use the command **return** to stop execution of the M-file without running any further commands. (The use of **return** here is a matter of style; we could instead have indented all of the remaining commands and put them between **else** and **end** statements.) If, on the other hand, **x** has some zero elements, then the **if** statement is false and the M-file skips ahead to the commands after the next **end** statement.

Often you can avoid the use of loops and branching commands entirely by using logical arrays. Here is another function M-file that performs the same task as in the previous examples. It has the advantage of being more concise and more efficient to run than the previous M-files, since it avoids a loop in all cases:

```
function y = f(x)
y = ones(size(x));
n = (x ~= 0);
y(n) = sin(x(n))./x(n);
```

Here **n** is a logical array of the same size as **x**, with a **1** in each place where **x** has a non-zero element and zeros elsewhere. Thus, the line that defines **y(n)** only redefines the elements of **y** corresponding to non-zero values of **x**, and leaves the other elements equal to **1**.

Branching with `switch`

The other main branching command is **switch**. It allows you to branch among several cases just as easily as among two cases, though the cases must be described through equalities rather than inequalities. Here is a simple example, which distinguishes between three cases for the input:

```
function y = count(x)
switch x
    case 1
        y = 'one';
    case 2
        y = 'two';
    otherwise
        y = 'many';
end
```

Here the **switch** statement evaluates the input **x**, and then execution of the M-file skips to whichever **case** statement has the same value. Thus, if the input **x** equals **1**, then the output **y** is set to be the string **'one'**, whereas if **x** is **2**, then **y** is set to **'two'**. In each case, once MATLAB encounters another **case** statement or an **otherwise** statement, it skips to the **end** statement, so that at most one case is executed. If no match is found among the **case** statements, then MATLAB skips to the (optional) **otherwise** statement, or else to the **end** statement. In the example above, an **otherwise** statement is present, so the output is **'many'** if the input is not **1** or **2**.

Unlike **if**, the command **switch** does not allow vector expressions, but it does allow strings. Type **help switch** to see an example using strings. This feature can be useful if you want to design a function M-file that uses a string input argument to select among several different variants of a program you write.

✓   Though strings cannot be compared with relational operators such as **==** (unless they happen to have the same length), you can compare strings in an **if** or **elseif** statement by using either **strcmp** or **isequal**. The latter command works more generally when comparing arrays that may have different sizes or even different data types.

More about Loops

In Chapter 3, we introduced the command **for**, which begins a loop – a sequence of commands to be executed multiple times. When you use **for**, you effectively specify the number of times to run the loop in advance (though this number may depend for instance on the input to a function M-file). Sometimes you may want to keep running the commands in a loop until a certain condition is met, without deciding in advance on the number of iterations. In MATLAB, the command that allows you to do so is **while**.

⇨   **Using while, you can easily end up accidentally creating an "infinite loop," one that will keep running indefinitely because the condition you set is never met. Remember that you can generally interrupt the execution of such a loop by typing CTRL+C; otherwise, you may have to shut down MATLAB.**

Open-Ended Loops

Here is a simple example of a script M-file that uses **while** to numerically sum the infinite series $1/1^4 + 1/2^4 + 1/3^4 + \cdots$, stopping only when the terms become so small (compared with the machine precision) that the numerical sum stops changing:

```
n = 1;
oldsum = -1;
newsum = 0;
while newsum > oldsum
    oldsum = newsum;
    newsum = newsum + n^(-4);
    n = n + 1;
end
newsum
```

Here we initialize **newsum** to **0** and **n** to **1**, and in the loop we successively add **n^(-4)** to **newsum**, add **1** to **n**, and repeat. The purpose of the variable **oldsum** is to keep track of how much **newsum** changes from one iteration to the next. Each time MATLAB reaches the end of the loop, it starts over again at the **while** statement. If **newsum** exceeds **oldsum**, the expression in the **while** statement is true, and the loop is executed again. But the first time the expression is false, which will happen when **newsum** and **oldsum** are equal, MATLAB skips to the **end** statement and executes the next line, which displays the final value of **newsum** (the result is **1.0823** to five significant digits). The initial value of **-1** that we gave to **oldsum** is somewhat arbitrary, but it must be negative so that the first time the **while** statement is executed, the expression therein is true; if we set **oldsum** to **0** initially, then MATLAB would skip to the **end** statement without ever running the commands in the loop.

✓ Even though you can construct an M-file like the one above without deciding exactly how many times to run the loop, it may be useful to consider roughly how many times it will need to run. Since the floating-point computations on most computers are accurate to about 16 decimal digits, the loop above should run until **n^(-4)** is about **10^(-16)**; that is, until **n** is about **10^4**. Thus the computation will take very little time on most computers. However, if the exponent were **2**, not **4**, the computation would take about **10^8** operations, which would take a long time on most (current) computers – long enough to make it wiser for you to find a more efficient way to sum the series, for example using **symsum** if you have the Symbolic Math Toolbox!

☞ *Though we have classified it here as a looping command,* **while** *also has features of a branching command. Indeed, the types of expressions allowed and the method of evaluation for a* **while** *statement are exactly the same as for an* **if** *statement. See the section* Logical Expressions *above for a discussion of the possible expressions you can put in a* **while** *statement.*

Breaking from a Loop

Sometimes you may want MATLAB to jump out of a **for** loop prematurely, for example if a certain condition is met. Or, in a **while** loop, there may be an auxiliary condition that you want to check in addition to the main condition in the **while** statement. Inside either type of loop, you can use the command **break** to tell MATLAB to stop running the loop and skip to the next line after the end of the loop. The command **break** is generally used in conjunction with an **if** statement. The following script M-file computes the same sum as in the previous example, except that it places an explicit upper limit on the number of iterations:

```
newsum = 0;
for n = 1:100000
    oldsum = newsum;
    newsum = newsum + n^(-4);
    if newsum == oldsum
        break
    end
end
newsum
```

In this example, the loop stops after **n** reaches **100000** or when the variable **newsum** stops changing, whichever comes first. In fact, **break** is involved as you can see by typing **n** after running the original version (using **while**) – it takes fewer than 10,000 iterations for **newsum** to equal **oldsum**. Notice that **break** ignores the **end** statement associated with **if** and skips ahead past the nearest **end** statement associated with a loop command, in this case **for**.

Other Programming Commands

Here we describe a few more advanced programming commands and techniques.

Subfunctions

In addition to appearing on the first line of a function M-file, the command **function** can be used later in the M-file to define an auxiliary function, or *subfunction*, which can be used anywhere within the M-file, but will not be accessible directly from the command line. For example, the following M-file sums the cube roots of a vector **x** of real numbers:

```
function y = sumcuberoots(x)
y = sum(cuberoot(x));

% ---- Subfunction starts here.
function z = cuberoot(x)
z = sign(x).*abs(x).^(1/3);
```

Here the subfunction **cuberoot** takes the cube root of **x** element-by-element. You can use subfunctions only in a function M-file, not in a script M-file. For examples of the use of subfunctions, you can examine many of MATLAB's built-in function M-files. For example, typing **type ezplot** will display four different subfunctions.

Cell and Structure Arrays

In Chapter 4, we discussed several data types, and earlier in this chapter we introduced another type, namely logical arrays. Two other data types that are useful in MATLAB programming are *cell arrays* and *structure arrays*. Cell arrays are essentially "arrays of arrays"; the elements of a cell array can have different data types and different sizes. The command **cell** creates an empty cell array, but the more common way to create a cell array is with curly braces:

```
>> ca = {1, [2 3], 'four'}

ca =
    [1]    [1x2 double]    'four'
```

Use curly braces also to access a particular element of a cell array; for example, type **ca{2}** to display the second element of **ca**.

Structure arrays are similar to structures in programming languages like C; they allow you to name the elements of the array rather than number them. Like cell arrays, the elements can have different types and sizes. One way to create a structure array is with the command **struct**:

```
>> sa = struct('data', [1 4 9 16 25], 'description', ...
'perfect squares')

sa =
           data: [1 4 9 16 25]
    description: 'perfect squares'
```

To access a particular element, or "field," type the name of the array and the name of the field with a period in between; for example, **sa.data**. You can also use the same syntax to create and add fields to a structure array. Another way to define **sa** is with the following commands:

```
>> sa = struct;
>> sa.data = [1 4 9 16 25];
>> sa.description = 'perfect squares'
```

The first command defines an empty structure, and the subsequent commands add fields to it.

✓ MATLAB also has signed and unsigned integer data types; see the online help for **int32** and **uint32**, for example.

✓ Cell and structural arrays are used primarily for the storage of, and ready access to, large data sets, as well as in sophisticated code generation. Another use is to establish a set of "options" to control MATLAB publishing. (See *Fine Points of Publishing* in Chapter 7.)

Commands for Parsing Input and Output

You may have noticed that many MATLAB functions allow you to vary the type and/or the number of arguments you give as input to the function. You can use the commands **nargin**, **nargout**, **varargin**, and **varargout** in your own M-files to handle variable numbers of input and/or output arguments, whereas to treat different types of input arguments differently you can use commands such as **isnumeric**, **ischar**, etc.

When a function M-file is executed, the functions **nargin** and **nargout** report, respectively, the number of input and output arguments that were specified on the command line. To illustrate the use of **nargin**, consider the following M-file add.m that adds either two or three inputs:

```
function s = add(x, y, z)
if nargin < 2
    error('At least two input arguments are required.')
end
if nargin == 2
    s = x + y;
else
    s = x + y + z;
end
```

First the M-file checks to see whether fewer than two input arguments were given, and if so it prints an error message and quits. (See the next section for more about **error** and related commands.) MATLAB automatically checks to see whether there are more arguments than specified on the first line of the M-file, so there is no need to do so within the M-file. If the M-file reaches the second **if** statement in the M-file above, we know that there are either two or three input arguments; the **if** statement selects the proper course of action in either case. If you type, for instance, **add(4,5)** at the command line, then, within the M-file, **x** is set to **4**, **y** is set to **5**, and **z** is left undefined; thus it is important to use **nargin** to avoid referring to **z** in cases where it is undefined.

To allow a greater number of possible inputs to add.m, we could add additional arguments on the first line of the M-file and add more cases for **nargin**. A better way to do this is to use the specially named input argument **varargin**.

```
function s = add(varargin)
s = sum([varargin{:}]);
```

In this example, all of the input arguments are assigned to the *cell array* **varargin**. The expression **varargin{:}** forms a comma-separated list of the input arguments. In the example above, we convert this list to a vector by enclosing it in square brackets, forming suitable input for **sum**.

The sample M-files above assume that their input arguments are numerical, and will attempt to add them even if they are not. This may be desirable in some cases; for instance, both M-files above will correctly add a mixture of numerical and symbolic inputs. However, if some of the input arguments are strings, the result will be either an essentially meaningless numerical answer or an error message that may be difficult to decipher. MATLAB has a number of test functions that you can use to make an M-file treat different types of input arguments differently – either to perform different calculations, or to produce a helpful error message if an input is of an unexpected type. For a list of some of these test functions, look up the commands beginning with **is** in the *Programming Commands* section of the Glossary.

As an example, here we use **isnumeric** in the M-file add.m to print an error message if any of the inputs are not numerical:

```
function s = add(varargin)
if ~isnumeric([varargin{:}])
  error('Inputs must be numerical: integer or floating-point.')
end
s = sum([varargin{:}]);
```

When a function M-file allows multiple output arguments, then, if fewer output arguments are specified when the function is called, the remaining outputs are simply not assigned. Recall that if no output arguments are explicitly specified on the command line, then a single output is returned and assigned to the variable **ans**. For example, consider the following M-file rectangular.m that changes coordinates from polar to rectangular:

```
function [x, y] = rectangular(r, theta)
x = r.*cos(theta);
y = r.*sin(theta);
```

Typing **[x, y] = rectangular(2, 1)** at the command line then stores the rectangular coordinates of the point with polar coordinates $(2, 1)$ in the variables **x** and **y**. But if you type only **rectangular(2, 1)**, then the answer will be just the x-coordinate. The following modification to **rectangular.m** adjusts the output in this case to be a complex number $x + iy$ containing both coordinates:

```
function [x, y] = rectangular(r, theta)
x = r.*cos(theta);
y = r.*sin(theta);
if nargout < 2
    x = x + i*y;
end
```

See the online help for **varargout** and the functions described above for additional information and examples.

Evaluation and Function Handles

The command **eval** allows you to run a command that is stored in a string as if you had typed the string on the command line. For example, typing **eval('cos(1)')** will produce the same result as typing **cos(1)**. If the entire command you want to run is contained in a string **str**, then you can execute it with **eval(str)**. Generally, **eval** is used in an M-file to define a variable or run a command whose name depends on an input or loop variable; type **help eval** for examples.

Another useful feature of **eval** is that, with two input strings, it will try the first command and, if that produces an error, try the second command instead. This form of **eval** allows some flexibility in the number of inputs to an anonymous function. The command

```
>> add = @(x,y,z) eval('x + y + z', 'x + y')
```

creates an anonymous function that adds either two or three input arguments, like add.m above.

✓ You can also create anonymous functions with multiple outputs using the command **deal**. For example, typing

```
>> rectangular = @(r,theta) deal(r.*cos(theta), ...
r.*sin(theta))
```

yields a function that converts a pair of polar coordinates into a pair of rectangular coordinates, like rectangular.m above. The only difference is that this function will yield an error message if you specify only one output argument. To allow for either one or two output arguments, you can use the two-input form of **eval** as in the example above; we leave the details to you as an exercise.

In Chapter 4, we mentioned function handles, which are now the preferred way to specify a function as input to another function (like **fzero** or **integral**, for example), though for backward compatibility many such functions also allow you to input a function name as a string. Function handles also make it simple for you to write function M-files that input another function. For example, the following M-file iterate.m inputs a function handle (or an inline function) and an initial value and iterates the function a specified number of times:

```
function final = iterate(func, init, num)
final = init;
for k = 1:num
    final = func(final);
end
```

Recall that putting the "at sign" **@** before the name of a built-in function or function M-file makes a handle for that function. Thus, typing **iterate(@cos, 1, 2)** yields the numerical value of $\cos(\cos(1))$, while **iterate(@cos, 1, 100)** yields an approximation to the real number x for which $\cos(x) = x$. (Think about it!) Now remember that an anonymous function is also a function handle. Therefore, it is the case that typing **iterate(@(x) 3.9*x*(1 - x), 0.5, 100)** iterates the "logistic map" $f(x) = 3.9x(1 - x)$ one hundred times starting at $x = 0.5$.

☞   *See* Population Dynamics *in the* Applications *chapter for more about the logistic map.*

Another feature of anonymous functions that is useful in programming is that their definition can include a parameter stored in a variable. (This is not the case for inline functions.) Consider the following commands:

```
>> c = 3; f = @(x) c*x; f(2)

ans =
    6
```

The value of **c** at the time **f** was defined is incorporated into **f**, and any subsequent changes to **c** do not affect **f**:

```
>> c = 5; f(2)

ans =
    6
```

This feature can be used in an M-file to define an anonymous function that depends on an input parameter. Another use is to vary a parameter in a loop. For example, the loop

```
>> hold on
>> for c = 1:3
     ezplot(@(x) sin(c*x), [0 2*pi])
end
>> hold off
```

will reproduce the curves in Figure 3.1.

User Input and Screen Output

In the previous section, we used **error** to print a message to the screen and then terminate execution of an M-file. You can also print messages to the screen without stopping execution of an M-file by using **disp** or **warning**. Not surprisingly, **warning** is intended to be used for warning messages, when the M-file detects a problem that might affect the validity of its result but is not necessarily serious. You can suppress warning messages, either from the command prompt or within an M-file, with the command **warning off all**. There are several other options for handling warning messages; type **help warning** for details.

In Chapter 4, we used **disp** to display the output of a command without printing the "ans =" line. You can also use **disp** to display informational messages on the screen while an M-file is running, or to combine numerical output with a message on the same line. For example, the commands

```
x = 2 + 2; disp(['The answer is ' num2str(x) '.'])
```

will set **x** equal to **4** and then print The answer is 4.

MATLAB also has several commands that solicit input from the user running an M-file. At the end of Chapter 3, we discussed three of them: **pause**, **keyboard**, and **input**. Briefly, **pause** simply pauses execution of an M-file until the user hits a key, while **keyboard** both pauses and gives the user a prompt that can be used like

the regular command-line interface before typing **return** to continue executing the
M-file. Lastly, **input** displays a message and allows the user to enter input for the
program on a single line. For example, in a program that makes successive approximations to an answer until some accuracy goal is met, you could add the following
lines to be executed after a large number of steps have been taken:

```
answer = input(['Algorithm is converging slowly; ', ...
     'continue (yes/no)? '], 's');
if ~isequal(answer, 'yes')
    return
end
```

Here the second argument **'s'** to **input** directs MATLAB not to evaluate the answer
typed by the user, just to assign it as a character string to the variable **answer**. We
use **isequal** to compare the answer to the string **'yes'** because **==** can be used
only to compare arrays (in this case, strings) of the same length. In this example,
we decided that if the user types anything but the full word **yes**, the M-file should
terminate. Other approaches would be to compare only the first letter **answer(1)**
with **'y'**, or to stop only if the answer is **'no'**, etc.

    If a figure window is open, you can use **ginput** to get the coordinates of a point
that the user selects with the mouse. As an example, the following M-file prints an
"X" where the user clicks:

```
function xmarksthespot
if isempty(get(0, 'CurrentFigure'))
    error('No current figure.')
end
flag = ~ishold;
if flag
    hold on
end
disp('Click on the point where you want to plot an X.')
[x, y] = ginput(1);
plot(x, y, 'xk')
if flag
    hold off
end
```

First, the M-file checks to see whether there is a current figure window. If so, it
proceeds to set the variable **flag** to **1** if **hold off** is in effect and **0** if **hold on**
is in effect. The reason for this is that we need to have **hold on** in effect to plot an
"X" without erasing the figure, but afterward we want to restore the figure window
to whichever state it was in before the M-file was executed. The M-file then displays
a message telling the user what to do, gets the coordinates of the point selected with
ginput(1), and plots a black "X" at those coordinates. The argument **1** to **ginput**
means to get the coordinates of a single point; using **ginput** with no input argument
would collect coordinates of several points, stopping only when the user presses the
ENTER key.

☞   *In Chapter 10, we describe how to create a Graphical User Interface (GUI)
within MATLAB to allow more sophisticated user interaction.*

Debugging

In Chapter 3, we discussed some rudimentary debugging procedures. One suggestion was to insert the command **keyboard** into an M-file, for instance right before the line where an error occurs, so that you can examine the Workspace of the M-file at that point in its execution. A more effective and flexible way to do this kind of debugging is to use **dbstop** and related commands. With **dbstop**, you can set a *breakpoint* in an M-file in a number of ways; for example, at a specific line number, or whenever an error occurs. Type **help dbstop** for a list of available options.

When a breakpoint is reached, a prompt beginning with the letter **K** will appear in the Command Window, just as if **keyboard** were inserted into the M-file at the breakpoint. In addition, the location of the breakpoint is highlighted with an arrow in the Editor/Debugger (which is opened automatically if you were not already editing the M-file). At this point, you can examine in the Command Window the variables used in the M-file, set another breakpoint with **dbstop**, clear breakpoints with **dbclear**, etc. If you are ready to continue running the M-file, type **dbcont** to continue or **dbstep** to step through the file line-by-line. You can also stop execution of the M-file and return immediately to the usual command prompt with **dbquit**.

☞   *You can also perform all the command-line functions that we described in this section with the mouse and/or keyboard shortcuts in the Editor/Debugger. See the section* Debugging Techniques *in Chapter 12 for more about debugging commands and features of the Editor/Debugger.*

☆ Interacting with the Operating System

☞   *This section is somewhat advanced. On a first reading, you might want to skip ahead to the next chapter.*

Calling External Programs

MATLAB allows you to run other programs on your computer from its command line. For example, you can conveniently enter UNIX or DOS file-manipulation commands directly in the Command Window rather than opening a separate window. Or, you may want to use MATLAB to graph the output of a program written in a language like Fortran or C. Finally, for large-scale computations, you can combine routines written in another programming language with routines you write in MATLAB.

The simplest way to run an external program is to type an exclamation point at the beginning of a line, followed by the operating-system command you want to run. For example, typing **!dir** on a Windows system or **!ls -l** on a UNIX system will generate a more detailed listing of the files in the current working directory than the MATLAB command **dir**. In Chapter 3, we described **dir** and other MATLAB commands like **cd**, **delete**, **pwd**, and **type** that mimic similar commands from the UNIX or DOS prompt. However, for certain operations you may need to run an appropriate command from the operating system.

✓ If you use the operating-system interface in an M-file that you want to run on either a Windows or UNIX system, you should use the test functions **ispc** and/or **isunix** to set off the appropriate commands for each type of system, e.g., **if isunix; ...; else; ...; end**. If you need to distinguish between different versions of UNIX (Linux and Mac OS), you can use **computer** instead of **isunix**.

The output from an operating-system command preceded by **!** can only be displayed to the screen. To assign the output of an operating-system command to a variable, you must use **dos** or **unix**. Though each is documented to work only for its respective operating system, in current versions of MATLAB they work interchangeably. For example, if you type **[stat, data] = dos('myprog 0.5 1000')**, the program **myprog** will be run with command-line arguments **0.5** and **1000**, and its "standard output" (which would normally appear on the screen) will be saved as a string in the variable **data**. (The variable **stat** will contain the exit status of the program you run, normally **0** if the program runs without error.) If the output of your program consists only of numbers, then **str2num(data)** will yield a row vector containing those numbers. You can also use **sscanf** to extract numbers from the string **data**; type **help sscanf** for details.

⇨ **A program you run with !, dos, or unix must be in the current directory or elsewhere in the path your system searches for executable files; the MATLAB path will not be searched.**

✓ You can access the program Perl, which is included in the MATLAB software. This is paticularly useful for string manipulation. See the online help for **perl** for more information.

✓ MATLAB also allows for more sophisticated interfaces with C, Fortran, and Java programs, but these are beyond the scope of this book. See the section on External Interfaces in the Help Browser for details.

File Input and Output

In Chapter 3, we discussed how to use **save** and **load** to transfer variables between the Workspace and a disk file. By default, the variables are written and read in MATLAB's own binary format, which is signified by the file extension .mat.

You can also read and write text files, which can be useful for sharing data with other programs. With **save**, type **-ascii** after the file name to save numbers as text rounded to eight digits, or **-ascii -double** for 16-digit accuracy. With **load**, the data is assumed to be in text format if the file name does not end in .mat. This provides an alternative to importing data with **dos** or **unix** in case you have previously run an external program and saved the results in a file. An even better alternative is to make use of the **Import Data** button ⤓ in the VARIABLE section of the HOME tab in the *Toolstrip*.

✓   MATLAB also offers an interactive tool called the *Import Wizard* to read data
        from files in different formats; to start it, type **uiimport** (optionally fol-
        lowed by a file name), select **Import Data...** in the VARIABLE section, or
        right-click on the file name in the Current Folder Browser and select **Import
        Data...**. Using **uiimport** also allows you to import data that you have cut
        or copied to the system clipboard.

        For more control over file input and output – for example to annotate numerical
output with text – you can use **fopen**, **fprintf**, and related commands. Type **help
iofun** for an overview of input and output functions.

Chapter 7

Publishing and M-Books

MATLAB is exceptionally strong in linear algebra, numerical methods, and graphical interpretation of data. It is easily programmed and relatively easy to learn to use. Hence, it has proven invaluable to engineers and scientists who rely on the scientific techniques and methods at which MATLAB excels. Very often the individuals and groups that employ MATLAB are primarily interested in the numbers and graphs that emerge from MATLAB commands, processes, and programs. Therefore, it is enough for them to work in a MATLAB Command Window, from which they can easily print or export their desired output.

However, other practitioners of mathematical software find themselves with two additional requirements. First, they need a mathematical software package embedded in an interactive environment, in which it is easy to make changes and regenerate results. Second, they need a higher-level presentation mode, which integrates computation and graphics with text, uses different formats for input and output, and communicates effortlessly with other software applications. These additional requirements can be accomplished either by editing and publishing an M-file, or else using the *M-book* interface, both of which were briefly described in Chapter 3. The present chapter goes into more detail and discusses some of the fine points of these methods.

Fine Points of Publishing

As we mentioned in Chapter 3, the simplest way to produce a finished presentation with MATLAB is to prepare your work in a script M-file that is divided into cells, and then publish the result. Publishing is most conveniently done in the Editor/Debugger. When you are ready to try publishing an M-file you have edited, click on the PUBLISH tab and then click on the button with the green triangle in the PUBLISH section of the Toolstrip. Your M-file will be saved automatically before being published. You can also use the command **publish** on a previously saved M-file.

There are a variety of output options that you can choose by selecting PUBLISH:PUBLISH:**Publish:Edit Publishing Options...**. By default, the output (regardless of file format) is saved in a folder named html. The default output format, HTML (which stands for "hypertext markup language," the formatting language for web pages), is probably most convenient for viewing the published output as you write and edit the M-file. MATLAB has a built-in web browser (which can be accessed directly with the command **web**) that will display your published document as soon as MATLAB has finishing processing it, and print the output if you wish. You can also put the HTML output on the internet by copying it to the appropriate location for your web server. Note that if you have any graphical output, MATLAB stores the results in

multiple files: one .html file with text input/output and comments, and one .png
file for each figure. All of these files must be copied in order to transmit the published
document. For this reason, you may prefer to save your final published document as a
single PDF file, which also can be put on the internet, or emailed to a colleague. Other
options include XML, which can be developed into a web page or document using
a variety of tools, and LATEX, the typesetting language that we used to produce this
book – in particular, Chapter 11 and the Solutions to the Practice Sets were published
to LATEX from M-files. On a Windows computer, you have two additional options –
Microsoft Word (with the file extension .doc) or, for presentation slides, PowerPoint
(with the file extension .ppt).

✓   The online help gives many examples of publishing features, but it is some-
      what vague about the general rules for how the input M-file determines what
      appears in the published output. Below we describe how publishing works
      based on our experience in 2013.

Sectioning

To produce a nicely formatted published document, recall from the section *Publishing
an M-file* in Chapter 3 that each new cell in the M-file starts a new section. Recall
also that a cell begins with a double percent sign %% at the start of a line, optionally
followed by a space and a section title. If you instead use a triple percent sign %%%,
you start a new section but do not start a new cell. Each section in the published output
contains four parts, each of which may be empty, in the following order:

1. Section title

2. Formatted text

3. MATLAB input

4. MATLAB output

    The title of the first section is used as the title of the published document, but only
if it appears on the first line of M-file, and if no MATLAB input appears in the first
section.

    If the section title is omitted, the published output will continue under the same
section heading as the previous section; while it is a separate section for the output
ordering rules, it will appear to the reader to be part of the same section. Untitled
sections also do not appear in the table of contents at the beginning of the published
document.

    Text that you want to be formatted should be a group of lines that each start with
a single percent sign %, separated by a space from the text that follows. This group
must appear immediately after the line that begins the section, without any intervening
blank lines. In the next section, we discuss a variety of options for how the text is
formatted.

    Starting with the first line in a section that does not begin with a percent sign,
each line until the end of the section is treated as MATLAB input and is reproduced
verbatim in the published output. For the rest of the section, comments starting with

% are not formatted, even if they begin a line. Depending on the output file format, MATLAB input may be enclosed in a box and may be color coded.

If MATLAB input is present, then any text output it produces (which would otherwise appear in the Command Window) is inserted into the published document, followed by any output that appears in a figure window. Graphics output comes after text output, even if the graphics input appears first. In addition, though all the text output appears, published graphics output includes only the contents of the figure window(s) after all the input in the section has been processed. (To avoid duplication, only figure windows that are changed or newly opened since the previous section are included.) You can use the command **snapnow** to force the contents of a figure window to be written to the published document before processing of the section has completed; this will also put the contents before any text output generated later in the section.

✓ There are several options for how the contents of figure windows are formatted in published output; see the item "Figure capture method" in PUBLISH:PUBLISH:**Publish**:**Edit Publishing Options...**. We suggest that you try the different options and see which you like best for your preferred output file format.

In order that the reader can understand which input line produced which output, we recommend that you start a new section (generally untitled) after each MATLAB command that you expect to produce text output. However, graphs are often produced with a sequence of commands (using commands like **hold**, **title**, and **axis**); to avoid intermediate graphs in the published output, include all the commands necessary to produce the finished graph in a single section. As a simple example, the code that produced the published solution to Problem 1(c) in Practice Set A reads as follows:

```
%% (c)
[2709/1024; 10583/4000; 2024/765]
%%
sqrt(7)
%%
% The third, that is 2024/765, is the best approximation.
```

The first section has a section title and MATLAB input but no formatted text, the second section has MATLAB input only, and the third section has formatted text only. The published solution has MATLAB output in each section with input, and appears to be a single section with title "(c)".

Formatted Text

As we discussed above, formatted text comes from a group of lines that start with a single percent sign, immediately following the double percent sign line that starts a section. Be careful to put a space between the percent sign and any text that follows on the line, or else formatting will stop with that line. Lines with a percent sign but no following text can be used to separate "blocks" that are to be formatted differently. Here are instructions for some of the available block formats:

- For text formatted into paragraphs, start with a percent sign **%**, then a single space, then text. To start a new paragraph, start a new block or a new section.

- For "preformatted" text, start with a percent sign **%** followed by two spaces. The rest of the block will be included verbatim, including the line breaks you choose, in a font similar to MATLAB input and output. If you start with a percent sign and three spaces, the formatting will be like MATLAB input in the Editor/Debugger, including coloring for some output file formats.

- For a bulleted list, start with a percent sign **%**, a space, an asterisk *****, and another space. Each line that starts with the same four-character sequence starts a new bullet. For a numbered list, replace the asterisk with a number sign **#**.

- For a mathematical formula displayed on its own line, start with a percent sign **%**, a space, and a double dollar sign **$$**. Also, end the line or block with a double dollar sign. In between, typeset the formula with LaTeX math mode commands, described below.

For each type of formatted text mentioned above, except the ordinary paragraph, you can insert a template block in the Editor/Debugger using one of the buttons in the PUBLISH:INSERT BLOCK MARKUP section.

There are also several formatting options within a paragraph or list block; buttons to produce templates for the main options are in the PUBLISH:INSERT INLINE MARKUP section. Enclosing text between a pair of asterisks ***** makes the text boldface, enclosing it between underscores **_** makes the text italic, and enclosing it between vertical bars **|** puts the text in the monospaced MATLAB input/output font. You can also add hyperlinks between a "less than" sign **<** and a "greater than" sign **>**. Finally, you can add symbols and formulas by typing appropriate commands between dollar signs **$**. The commands are taken from the math mode syntax of LaTeX; we give a few examples here. You can create a Greek letter by typing the name of the letter preceded by a backslash ****; capitalize the name of the letter to get a capital Greek letter. You can create a subscript or superscript by preceding it with an underscore **_** or caret **^**, respectively. Grouping is done with braces **{ }**, for example to enclose a multi-character subscript or superscript. Mathematical symbols have names (usually self-explanatory) beginning with a backslash ****, such as **\times** for a multiplication sign and **\sum** for a summation sign. Here is an example of an M-file suitable for publishing, using some of the formatting techniques described above:

```
%% Sample M-File About Bessel Functions
% The _Bessel function_ J_n(x) can be defined in many ways:
%
% * For example, for |n| an integer, J_n(x) can be defined
% by the formal series expansion
%
% $$ e^{z(t-1/t)/2} = \sum_{n=-\infty}^\infty t^n J_n(z). $$
%
% * Secondly, J_n(x) can be defined as the solution of
%
% $$ x^2y''(x) + xy'(x) + (x^2 - n^2)y(x)=0 $$
%
% that is non-singular at x = 0 and satisfies a
% normalization condition.
%%
```

```
% Here is a graph of some of the Bessel functions:
x = 0:0.01:10;
for n = 0:5
    plot(x, besselj(n, x)), hold on
end, hold off
```

If you publish this example to LaTeX, the LaTeX code for the equations will be incorporated verbatim. If you publish to another file format, MATLAB typesets each equation and stores the result as a graphics file. Depending on the file format, the graphics file may be embedded into the document file or stored as a separate file, just as with MATLAB graphics output.

Other Tips

You may find (in particular, for publishing to PDF) that the default size for graphics in a published document is too large for your purposes. You can set a maximum size by selecting PUBLISH:PUBLISH:**Publish**:**Edit Publishing Options...** in the Editor/Debugger and using the items "Maximum Image Height" and/or "Maximum Image Width". We find that setting the maximum width to around 350 pixels produces reasonably sized figures in PDF output. The actual size of a graphic may be smaller, depending on the size of the figure window that produced it.

Recall that publishing works only with script M-files, not with function M-files. If you are doing a calculation that requires a function M-file and you want to publish your results, you can call the function M-file from a publishable script M-file, and include in the latter the line

```
type <function M-file>
```

where **<function M-file>** should be replaced with the name of the M-file. This will ensure that the text of the function M-file appears in your published document. You can see examples of this technique in the section *Numerical Solution of the Heat Equation* in Chapter 11.

One problem you may encounter occurs when you publish an M-file that has a mistake in it, or at least that contains code that produces a MATLAB error message. Probably you will want to fix the error, but you may want to retain the erroneous command for illustrative purposes. (For an example, see the solution to Practice Set B, Problem 17, in *Solutions to the Practice Sets*.) Whether you publish or simply run an M-file, MATLAB terminates as soon as it encounters an error, and does not evaluate the rest of the M-file. A way around this is to use the commands **try** and **catch**. The input line

```
try; ezplot(1); end
```

tries the **ezplot** command and, on encountering an error, resumes execution of the M-file after the **end** statement. However, the error message is not printed. The following commands will display the error message and continue execution:

```
try
    ezplot(1)
catch err
    disp(err.message)
end
```

We conclude this section with one more fine point about what happens when you publish to LaTeX. If you do this, the output LaTeX code (a `.tex` file) and graphics files (`.eps`, for "Encapsulated PostScript") are stored in your `html` folder. You can immediately run `latex` on the `.tex` file, or else you can paste it into another LaTeX file. If you do the latter, you have to remember to remove the lines

`\documentclass{article}`

and

`\begin{document}`

at the beginning and

`\end{document}`

at the end, and you should move the lines

`\usepackage{graphicx}`
`\usepackage{color}`

to the preamble of the document into which you are pasting the code (unless they are already present there, in which case you should delete the extra copies). Of course, you can optionally alter or remove other parts of the output; we removed the tables of contents from the published output later in this book.

More on M-Books

The M-book interface allows a Windows user to operate MATLAB from a special Microsoft Word document instead of from a MATLAB Command Window. In this mode, the user should think of Word as running in the foreground and MATLAB as running in the background as an *Automation Server*. Lines that you enter into your Word document are passed to the MATLAB server in the background and executed there, whereupon the output is returned to Word, and then both input and output are automatically formatted. You obtain a living document, in the sense that you can edit the document as you would normally edit a word-processing document. So you can revisit input lines that need adjustment, change them, and re-execute them on the spot – after which the old outdated output is automatically overwritten with new output. The graphical output that results from MATLAB graphics commands appears in the Word document, immediately after the commands that generated it. Erroneous input and output are easily expunged, enhanced formatting can be done in a way that is no more complicated than word processing, and, in the end, the result of your MATLAB session can be an attractive, easily readable, and highly informative document. Of course, you can also "cheat" by editing your output – we shall discuss that and other pitfalls and strengths of M-books in what follows.

☞   *We are assuming here that M-books have been enabled on your computer. If not, or if you get an error message about a missing file* `m-book.dot`*, go back to the section on* M-Books *in Chapter 3 for instructions on how to fix this problem.*

The most common way to start up the M-book interface is to type **notebook** at the Command Window prompt, or, if you want to open an existing M-book, to type

```
>> notebook <M-book>
```

with **<M-book>** replaced by the name of the M-book, ordinarily ending in .docx. The M-book appears in a Word window, and you can enter text in the usual way. The MATLAB automation server is activated only if you do one of two things: either use the items in the **Notebook** menu (accessed by clicking the **Add-Ins** tab), or press the key combination CTRL+ENTER. You may note that your commands take a little longer to evaluate than they would inside a normal MATLAB Command Window. This is not surprising considering the amount of information that is passing back and forth between MATLAB and Word.

✓   If you want to start a fresh M-book, select **File:New** in Word, click on "My Templates..." in the window that appears, then select m-book. You can also open a previously saved M-book, either directly through **File:Open**, or by double-clicking on the file name in Windows Explorer. Word will automatically launch MATLAB if it is not running already.

The Notebook Menu Items

Next let's examine the items in the **Notebook** menu. First comes **Define Input Cell**. If you click on any line and select **Define Input Cell**, then that line will become an input line. To evaluate it, you must press CTRL+ENTER or choose **Notebook:Evaluate Cell**. You can also create an input cell containing more than one line. For example, type

```
syms x y
factor(x^2 - y^2)
```

then select both lines (by clicking and dragging over them) and choose **Define Input Cell**. Pressing CTRL+ENTER will then cause *both* lines to be evaluated. You can recognize that both lines are incorporated into one input cell by looking at the brackets, or *Cell Markers*. The menu item **Hide Cell Markers** will cause the Cell Markers to disappear; in fact that menu item is a toggle switch that turns the Markers on and off. If you have several input cells, you can convert them into one input cell by selecting them and choosing **Group Cells**. You can break them apart by choosing **Ungroup Cells**. If you click in an input cell and choose **Undefine Cells**, that cell ceases to be an input cell; its formatting reverts to the default Word format, as does the corresponding output cell. If you "undefine" an output cell, it loses its format, but the corresponding input cell remains unchanged.

If you select some portion of your M-book (e.g., the entire M-book by using **Select:Select All** in the Editing section of the Home tab) and then choose **Purge Selected Output Cells**, all output cells in the selection will be deleted. This is particularly useful if you wish to change some data on which the output in your selection depends, and then re-evaluate the entire selection by choosing **Evaluate Cell**. You can re-evaluate the entire M-book at any time by choosing **Evaluate M-book**. If your M-book contains a loop, you can evaluate it by selecting it and choosing **Evaluate Loop...**, or for that matter **Evaluate Cell**, provided that the entire loop is inside a single input cell.

It is often handy to purge all output from an M-book before saving, to economize on storage space or on time upon reopening, especially if there are complicated graphs in the document. If there are any input cells that you want to automatically evaluate upon opening of the M-book, select them and click on **Define AutoInit Cell**. The color of the text in those cells will change from green to navy blue. If you want to separate out a series of commands, say for repeated evaluation, then select the cells and click on **Define Calc Zone**. The commands selected will be encased in a Word section (with section breaks before and after it). If you click in the section and select **Evaluate Calc Zone**, the commands in only that zone will be (re)evaluated.

The last two buttons are also useful. The button **Bring MATLAB to Front** does exactly that; it reveals the MATLAB Command Window that has been hiding behind the M-book. You may want to enter some commands directly into the Command Window and not have them in your M-book – for example, a **help** entry. Finally, the last button, **Notebook Options...** brings up a panel in which you can do some customization of your M-book: set the numerical format, establish the size of graphics figures, etc.

M-Book Graphics

All MATLAB commands that generate graphics work in M-books. The figure produced by a graphics command appears immediately below that command. However, if you refine your graphics in an M-book, you will find that it is more desirable to modify the input cells that generated them, rather than producing more pictures by repeating the command with new options. So when adding things like **xlabel**, **ylabel**, **legend**, **title**, etc., it is usually best just to add them to the graphics input cell and re-evaluate.

In most cases, it is desirable to have figures appear directly in the M-book, using the default option **Notebook:Notebook Options...:Embed figures in MATLAB Notebook**. Figures will also appear in a MATLAB figure window, which allows manipulations not available in the M-book, such as rotating a 3D plot with the mouse. However, the results of these manipulations will not appear in the M-book.

Finally, we note the button **Toggle Graph Output for Cell**, the only button on the **Notebook** menu not previously described. If you select a cell containing a graphics command and click on this button, no graphical output will result from the evaluation of this command. This can be useful when used in conjunction with **hold on** if you want to produce a single graphic using multiple command lines.

More Hints for Effective Use of M-Books

You may want to run script or function M-files in an M-book. You must still make sure that these M-files are located in your MATLAB path, just as you would in a Command Window. But, assuming that this is the case, M-files are executed in an M-book exactly as in a Command Window. You invoke them simply by typing their names and pressing CTRL+ENTER. The outputs they generate, both intermediate and final, are determined as before. In particular, semicolons at ends of lines are important.

A standard MATLAB feature that does not work so well in M-books is the **. . .** construct for continuing a long command entry on a second line. Word automatically converts three dots into a single special *ellipsis* character and so confuses MATLAB. There are two ways around this difficulty. One way is not to use ellipses; rather, simply continue typing and allow Word to wrap as usual – the command will be interpreted properly when passed to MATLAB. Another method is to override the "AutoCorrect" feature of Word that converts the three dots into an ellipsis. This is most easily done by typing CTRL+Z after the three dots.

The ellipsis difficulty described in the last paragraph is just one of many of its type. The various kinds of automatic formatting that Word carries out can truly confuse MATLAB. Several such instances that we find particularly annoying are as follows: fractions (1/2 is converted to a single character $^1/_2$ representing one-half); the character combination ":)", a construct often used when specifying the rows of a matrix, which Word converts to a "smiley face" ☺; and various dashes that wreak havoc with MATLAB's attempts to interpret an ordinary hyphen as a minus sign. You can turn off such substitutions as follows: select **File:Options**, then in the resulting window click on "Proofing" followed by "AutoCorrect Options...", then change the settings that appear.

A more insidious problem is the following. If you copy and paste text into an existing input cell, the font may not match the usual input font even after evaluation, and the cell may not evaluate correctly. In general, you should copy only from another input cell that has already evaluated successfully – otherwise, it is safer to type in the line anew.

Finally, we have seen instances in which a cell, for no discernible reason, fails to evaluate. If this happens, try pressing CTRL+ENTER again. If all else fails, your only choice may be to delete and retype the cell.

Practice Set B

Math, Graphics, and Programming

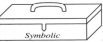 Problems 2, 3, 5–7, and parts of 10–12 require the Symbolic Math Toolbox. The others do not.

1. Use **contour** to do the following:

   (a) Plot the level curves of the function $f(x, y) = 3y + y^3 - x^3$ in the region where x and y are between -1 and 1 (to get an idea of what the curves look like near the origin), and in some larger regions (to get the big picture).

   (b) Plot the curve $3y + y^3 - x^3 = 5$.

   (c) Plot the level curve of the function $f(x, y) = y \ln x + x \ln y$ that contains the point $(1, 1)$.

2. Find the derivatives of the following functions, and if possible, simplify each answer:

   (a) $f(x) = \dfrac{2x - 1}{x^2 + 1}$

   (b) $f(x) = \sin(3x^2 + 2)$

   (c) $f(x) = x^r$.

3. See whether MATLAB can do the following integrals symbolically, and for the indefinite integrals, check the results by differentiating:

   (a) $\int_0^{\pi/2} \cos x \, dx$

   (b) $\int \sin(3x) \sqrt{1 - \cos(3x)} \, dx$

   (c) $\int x^2 \sqrt{x + 3} \, dx$

   (d) $\int_{-\infty}^{\infty} e^{-x^2} \, dx$.

4. Compute the following integrals numerically using **integral**. When possible, find the error in the numerical answer by comparing it with the exact answer:

   (a) $\int_0^{\pi} e^{\sin x} dx$

(b) $\int_0^1 \sqrt{x^3 + 1}\, dx$

(c) $\int_{-\infty}^{\infty} e^{-x^2}\, dx$.

5. Evaluate the following limits:

(a) $\lim\limits_{x \to -\pi} \dfrac{1 + \cos x}{x + \pi}$

(b) $\lim\limits_{x \to \infty} x^2 e^{-x}$

(c) $\lim\limits_{x \to 1^-} \dfrac{1}{x - 1}$

(d) $\lim\limits_{x \to 0^+} \sin\left(\dfrac{1}{x}\right)$.

6. Compute the following sums:

(a) $\displaystyle\sum_{k=1}^{n} k^2$

(b) $\displaystyle\sum_{k=0}^{n} r^k$

(c) $\displaystyle\sum_{k=0}^{\infty} \dfrac{x^k}{k!}$.

7. Find the Taylor polynomial of the indicated degree n at the indicated point c for the following functions:

(a) $f(x) = e^x$, $n = 6$, $c = 0$

(b) $f(x) = \sin x$, $n = 5$, $c = 2$

(c) $f(x) = \ln x$, $n = 4$, $c = 1$.

8. Plot the following surfaces:

(a) $z = \sin x \sin y$ for $-3\pi \le x \le 3\pi$ and $-3\pi \le y \le 3\pi$

(b) $z = (x^2 + y^2) \cos(x^2 + y^2)$ for $-1 \le x \le 1$ and $-1 \le y \le 1$.

9. Create a 17-frame movie, whose frames show filled red circles of radius $1/2$ centered at the points $\left(4\cos(j\pi/8), 4\sin(j\pi/8)\right)$, $j = 0, 1, \ldots, 16$. Make sure that all the circles are drawn on the same set of axes, and that they look like circles, not ellipses.

10. In this problem, we use the *backslash* operator, or "left-matrix-divide" operator, introduced in the section *Solving Linear Systems* of Chapter 4.

    (a) Use the backslash operator to solve the system of linear equations in Problem 3 of Practice Set A.

    (b) Now try the same method on Problem 4 of Practice Set A. MATLAB finds one, but not all, answer(s). Can you explain why? If not, see Problem 11 below, as well as part (d) of this problem.

(c) Next try the method on this problem:

$$w + 3x - 2y + 4z = 1$$
$$-2w + 3x + 4y - z = 1$$
$$-4w - 3x + y + 2z = 1$$
$$2w + 3x - 4y + z = 1.$$

Check your answer by matrix multiplication.

(d) Finally, try the matrix division method on:

$$ax + by = u$$
$$cx + dy = v.$$

Don't forget to declare the variables to be symbolic. Your answer should involve a fraction, and so will be valid only when its denominator is non-zero. Evaluate **det** on the coefficient matrix of the system. Compare this with the denominator.

11. We deal in this problem with 3×3 matrices, although the concepts are valid in any dimension.

(a) Consider the rows of a square matrix A. They are vectors in 3-space and so span a subspace of dimension 3, 2, 1, or possibly 0 (if all the entries of A are zero). That number is called the *rank* of A. The MATLAB command **rank** computes the rank of a matrix. Try it on the four coefficient matrices in each of the parts of Problem 10. Comment on MATLAB's answer for the fourth one.

(b) An $n \times n$ matrix is *non-singular* if its rank is n. Which of the four you computed in part (a) are non-singular?

(c) Another measure of non-singularity is given by the *determinant* – a fundamental result in linear algebra is that a matrix is non-singular precisely when its determinant is non-zero. In that case, a unique matrix B exists that satisfies $AB = BA =$ the identity matrix. We denote this inverse matrix by A^{-1}. MATLAB can compute inverses with **inv**. Compute **det(A)** for the four coefficient matrices, and for the non-singular ones, find their inverses. Note: the matrix equation $Ax = b$ has a unique solution, namely $x = A^{-1}b = $ **A\b**, when A is non-singular.

12. As explained in Chapter 4, when you compute **[U, R] = eig(A)**, each column of U is an eigenvector of A associated with the eigenvalue that appears in the corresponding column of the diagonal matrix R. This says exactly that $AU = UR$.

(a) Verify the equality $AU = UR$ for each of the coefficient matrices in Problem 10.

(b) In fact, $\text{rank}(A) = \text{rank}(U)$, so, when A is non-singular,

$$U^{-1}AU = R.$$

Thus, if two non-singular matrices A and B have the same set of eigen-vectors, then the fact that diagonal matrices commute implies that A and B commute too. Verify these facts for the two matrices

$$A = \begin{pmatrix} 1 & 0 & 2 \\ -1 & 0 & 4 \\ -1 & -1 & 5 \end{pmatrix}, \quad B = \begin{pmatrix} 5 & 2 & -8 \\ 3 & 6 & -10 \\ 3 & 3 & -7 \end{pmatrix};$$

that is, show that the matrices of eigenvectors are the "same" – that is, the columns are the same up to a scalar multiple – and verify that $AB = BA$.

13. This problem, having to do with genetic inheritance, is based on Chapter 12 in C. Rorres and H. Anton, *Applications of Linear Algebra*, 3rd ed., Wiley, New York, 1984. In a typical inheritance model, a trait in the offspring is determined by the passing of a genotype from the parents, where there are two independent possibilities from each parent, say A and a, and each is equally likely. (A is the dominant gene, and a is recessive.) Then we have the following table of proba-bilities of the possible genotypes for the offspring for all possible combinations of the genotypes of the parents:

Genotype of Offspring	Parental Genotype					
	AA-AA	AA-Aa	AA-aa	Aa-Aa	Aa-aa	aa-aa
AA	1	1/2	0	1/4	0	0
Aa	0	1/2	1	1/2	1/2	0
aa	0	0	0	1/4	1/2	1

Now suppose that we have a population in which mating occurs only with an identical genotype. (That's not far-fetched if we are considering a controlled plant or vegetable population.) Next suppose that x_0, y_0, and z_0 denote the percentages of the population with genotype AA, Aa, and aa, respectively, at the outset of observation. We then denote by x_n, y_n, and z_n the percentages in the nth generation. We are interested in knowing these numbers for large n, and how they depend on the initial population. Clearly

$$x_n + y_n + z_n = 1, \quad n \geq 0.$$

Now we can use the table to express a relationship between generations n and $n + 1$. Because of our presumption on mating, only the first, fourth, and sixth columns are relevant. Indeed a moment's reflection reveals that we have

$$x_{n+1} = x_n + \frac{1}{4}y_n$$

$$y_{n+1} = \frac{1}{2}y_n$$

$$z_{n+1} = z_n + \frac{1}{4}y_n.$$

(a) Write these equations as a single matrix equation $X_{n+1} = MX_n$, $n \geq 0$. Explain carefully what the entries of the column matrix X_n are, and what the coefficients of the square matrix M are.

(b) Apply the matrix equation recursively to express X_n in terms of X_0 and powers of M.

(c) Next, use MATLAB to compute the eigenvalues and eigenvectors of M.

(d) From Problem 12, you know that $MU = UR$, where R is the diagonal matrix of eigenvalues of M. Solve that equation for M. Can you see what $R_\infty = \lim_{n \to \infty} R^n$ is? Use that and your above expression of M in terms of R to compute $M_\infty = \lim_{n \to \infty} M^n$.

(e) Describe the eventual population distribution by computing $M_\infty X_0$.

(f) Check your answer by directly computing M^n for large specific values of n. (*Hint*: MATLAB can compute the powers of a matrix **M** by entering **M^10**, for example.)

(g) You can alter the fundamental presumption in this problem by assuming, alternatively, that all members of the nth generation must mate only with a parent whose genotype is purely dominant. Compute the eventual population distribution of that model. Chapters 12–14 in Rorres and Anton have other interesting models.

14. ☆ The French flag is 1.5 times as wide as it is high, and is divided into three vertical stripes, colored (in order) blue, white, and red. The Italian flag is the same, except that blue is replaced by green. Create a $200 \times 300 \times 3$ array representing the French flag, view it in a figure window, and convert it to a JPEG file `tricolore.jpg`. Do the same with the Italian flag, this time converting to a file `italia.jpg`. Finally, create a movie showing the French flag "transform" into the Italian one.

15. For a positive integer n, let $A(n)$ be the $n \times n$ matrix whose entry in the (i, j)-position is $a_{ij} = 1/(i + j - 1)$. For example,

$$A(3) = \begin{pmatrix} 1 & \frac{1}{2} & \frac{1}{3} \\ \frac{1}{2} & \frac{1}{3} & \frac{1}{4} \\ \frac{1}{3} & \frac{1}{4} & \frac{1}{5} \end{pmatrix}.$$

The eigenvalues of $A(n)$ are all real numbers. Write a script M-file that prints the largest eigenvalue of $A(500)$, without any extraneous output. (*Hint*: the M-file may take a while to run if you use a loop within a loop to define A. Try to avoid this!)

16. ☆ Write a script M-file that draws a bulls-eye pattern with a central circle colored red, surrounded by alternating circular strips (annuli) of white and black, say ten of each. Make sure that the final display shows circles, not ellipses. (*Hint*: one way to color the region between two circles black is to color the entire inside of the outer circle black, then color the inside of the inner circle white.)

17. MATLAB has a function **lcm** that finds the least common multiple of two numbers. Write a function M-file `mylcm.m` that finds the least common multiple of an arbitrary number of positive integers, which may be given as separate arguments or in a vector. For example, **mylcm(4, 5, 6)** and **mylcm([4 5 6])** should both produce the answer **60**. The program should produce a helpful error message if any of the inputs are not positive integers. (*Hint*: for three numbers you could use **lcm** to find the least common multiple m of the first two numbers, then use **lcm** again to find the least common multiple of m and the third number. Your M-file can generalize this approach.)

18. ☆ Write a function M-file that takes as input a string containing the name of a text file and produces a histogram of the number of occurrences of each letter from A to Z in the file. Try to label the figure and axes as usefully as you can.

19. ☆ Write a function M-file (with no arguments or outputs) that scans the current folder for the most recently modified M-file and opens it in the Editor/Debugger. If the current folder contains no M-files, your M-file should produce an error message. Ideally, you should try to write the M-file so that it works in any operating system.

Chapter 8

MuPAD

In this chapter, we discuss the MuPAD language, which underlies MATLAB's Symbolic Math Toolbox. We believe that the most practical uses of MuPAD are to be found in calculus, linear algebra, number theory, combinatorics, and differential equations. We shall show how to use MuPAD *directly* to deal with issues from some of those topics. We shall comment more fully below on *why* you might want to do so. For additional basic information on MuPAD (beyond what you will find in this book), we recommend the MuPAD Tutorial, which you can find at the `mathworks.com` web site. (MathWorks is the software company that produces MATLAB and MuPAD.)

As we mentioned at the end of Chapter 4, all of the commands in the Symbolic Math Toolbox rely on a software package called MuPAD. Most of the time you may not need to know this, but for sophisticated symbolic calculations it helps to work directly in MuPAD, without going through MATLAB as an intermediary. A rather dramatic example of this is described in the authors' book *Differential Equations with MATLAB*, 3rd ed., Wiley, New York, 2012, where in the topic of series solutions of ordinary differential equations, it is far clumsier to work through MATLAB than it is to invoke MuPAD symbolic commands directly. The reader may well encounter other topics in calculus, algebra, or differential equations for which a direct application of MuPAD is more powerful, effective, and convenient than working through the MATLAB "translator." Furthermore, there will also be times where the Symbolic Math Toolbox gives mysterious or non-optimal results, and knowing something about MuPAD will enable you to understand why this happens and what you can do about it. (In this regard, see ibid., Chapter 6.)

The advantages of using MuPAD directly, rather than accessing the Symbolic Math Toolbox via MATLAB, is explained at some length on the MathWorks web site. At the time of this writing, the relevant URL on the MathWorks site is:

`www.mathworks.com/help/symbolic/computations-in-the-`
`    matlab-command-window-vs-the-mupad-notebook-interface.html`

But for brevity, here are the most salient points (actual direct quotes from the site), which we will address in the discussion later in this chapter.

- For some computations, the performance is better in the MuPAD notebook interface than in the MATLAB Command Window. The reason is that the engine returns the results in the MuPAD language. To display them in the MATLAB Command Window, the Symbolic Toolbox translates the results to the MATLAB language (sometimes less than perfectly).

- Working in the MuPAD notebook interface is recommended if you want to document your work and results, for example, embed graphics and descriptive text with your calculations. The MuPAD notebook interface displays results in type-set math, making them look very similar to what you see in mathematical books. Also, graphics in the MuPAD notebook interface are interactive, which, alas, we will not be able to illustrate here.

- MuPAD libraries support computations in: Combinatorics, Number Theory, Graph Theory, and many other topics listed on the site. MuPAD libraries also provide large collections of functions for working in these subjects.

- The MuPAD notebook interface provides many more options for modifying common symbolic functions than MATLAB provides.

Starting MuPAD

To start MuPAD by itself, simply type **mupad** from the MATLAB command prompt. This will open a MuPAD notebook that looks like the one shown in Figure 8.1.

Figure 8.1. A Blank MuPAD Notebook.

The MuPAD notebook acts very much like a stand-alone version of the MATLAB desktop. In the main MuPAD window, a small bracket will show a spot for typing commands. MuPAD commands appear in the notebook in red, with the output in blue just below, so that it's easy to tell them apart. At any time, you can edit one of the commands and hit ENTER to reevaluate it.

MuPAD *vis-à-vis* MATLAB

We encourage you to experiment with MuPAD on your own, but it helps to highlight a few crucial differences between MuPAD and MATLAB syntax:

- In MuPAD, all variables are by default symbolic. There is no need to "declare" symbolic variables as in MATLAB; this is quite convenient if you intend to do a lot of symbolic computations. The MuPAD command **float** can be used (instead of **double** in MATLAB) to convert a number to floating-point form.

- Certain basic operators behave differently in MuPAD and MATLAB. In MuPAD, **:=** (rather than **=**) is used for assignments, and **=** is used for equations. This takes some getting used to. Should you assign **w:=4**, for example, you can clear it by typing **delete w**.

- The MuPAD notebook comes with a Command Bar at the right. This makes it easy to format your commands and to recall the syntax for some basic functions.

- In a MuPAD notebook, you have the possibility of inserting headings and lines of explanatory text, just as you would in a published M-file. "Text areas" in the notebook are convenient for the reader, but are not evaluated by the symbolic engine. The simplest way to introduce text into a MuPAD notebook is via the **Text Paragraph** option in the **Insert** button on the menu bar. To go back to calculating, click on **Insert:Calculation**. There are also icons for these steps on the MuPAD tool bar.

- To suppress output of a command, terminate it with a colon (**:**), not a semicolon. You can separate commands within an input cell using commas or semicolons, but then the output for each command will be displayed. (Of course, this is often what you want.)

- In MuPAD, the "two dots" operator **..**, rather than a colon or a list enclosed in square brackets, is used to denote a range of values, either for creating a list or for giving the range of values in a plotting command. The special operator **\$** is used for generating sequences or lists. Thus **[x^2 \$ x = -10..10]** is the MuPAD substitute for the MATLAB construct **(-10:10).^2**.

- Sometimes you will want to type a long command on more than one line. In this case, the key combination ALT+ENTER within an "input cell" enables you to introduce a line break without having MuPAD evaluate what you've typed so far.

- In MuPAD, you access the second item in a list or array by appending **[2]**, not **(2)**, after the array or list name.

- In MuPAD, strings are delimited by double, not single, quote marks.

- To combine plots, use the **plot** command, instead of **hold on**.

- The syntax for the **subs** command is slightly different, and uses an "equals sign" (**=**) in place of the second comma.

- In MuPAD, you can use the percent sign (**%**) to refer to the output of the previously executed command. (We shall use that below.) However, in MATLAB you typically have to refer to **ans**, or the actual name if you assigned the output of the previous command to a variable.

As a simple first example, Figure 8.2 shows a MuPAD notebook with a heading and some text, and with commands to plot the parabola $y = x^2 + 5x - 3$ and to solve the quadratic equation $x^2 + 5x - 3 = 0$, both symbolically and numerically.

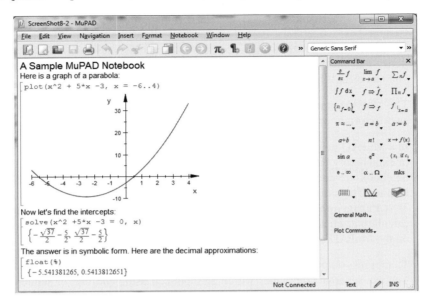

Figure 8.2. A MuPAD Notebook for Studying the Parabola $y = x^2 + 5x - 3$.

At this point, you could save the notebook, say as the file `parabola.mn` (the `.mn` extension stands for "MuPAD notebook"), using **Save As...** from the **File** menu at the upper left. You could also convert it to a PDF or HTML file (to show to a colleague, for example) with the **Export...** option of the **File** menu. This feature of MuPAD is a very convenient alternative to a published M-file when you want to produce a finished document combining computations, text, and graphics.

MuPAD has analogs of virtually all the graphics, algebraic, calculus, linear algebra, and basic structural commands discussed in the first five chapters of this book. Moreover, there are numerous options to these commands – exactly as there are in MATLAB – that help you to refine your input in order to render your output more meaningful. Since this book is not a MuPAD manual, we leave it to readers to explore these commands and options by reading the help and experimenting on your own. We shall satisfy ourselves by presenting some typical examples to illustrate the four bullets extracted from the MathWorks web site, presented at the beginning of this chapter.

The MuPAD Notebook Interface

Figures 8.3 to 8.5 depict a MuPAD notebook that contains a self-contained illustration of the *Mean Value Theorem*, which contains text, graphics, and mathematical computations in an integrated format.

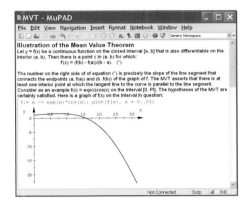

Figure 8.3. First page of the Mean Value Theorem Notebook.

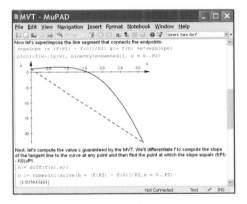

Figure 8.4. Second page of the Mean Value Theorem Notebook.

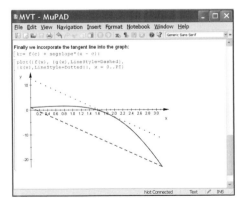

Figure 8.5. Third page of the Mean Value Theorem Notebook.

Additional Options to MATLAB Commands

Some of MATLAB's symbolic commands offer additional options in the MuPAD notebook interface. Here is a simple example:

```
>> syms x y; [x y] = solve(x^2 + y == 0, x + y == 0)
```

```
x =
 0
 1
y =
 0
 -1
```

But the option **VectorFormat**, which is not available in the MATLAB Command Window, can be invoked in MuPAD to yield:

```
solve({x^2 + y = 0, x + y = 0}, {x, y}, VectorFormat)
```

$$\left(\begin{array}{c} x \\ y \end{array} \right) \in \left\{ \left(\begin{array}{c} 0 \\ 0 \end{array} \right), \left(\begin{array}{c} 1 \\ -1 \end{array} \right) \right\}$$

MuPAD Libraries Support Specialized Topics

There are many functions and commands, addressed to specialized topics, that are available in MuPAD, but not in MATLAB. Some of these topics include: number theory, combinatorics, statistics, and many others. For example, you can do modular arithmetic directly at the command line in MuPAD, but not in MATLAB:

```
15 div 4
```

   3

```
13 mod 4
```

   1

Actually, the **mod** command (but not **div**) exists in MATLAB. But it will typically fail for large integers, when round-off errors can come into play. MuPAD has a more powerful command called **powermod**. Here we use it to illustrate *Fermat's Theorem*: If p is a prime number and $b \not\equiv 0 \pmod{p}$, then $b^{p-1} \equiv 1 \pmod{p}$. Here $b = 2$ and the prime is 10000000019:

```
powermod(2, 10000000018, 10000000019)
```

   1

You can learn about all of the MuPAD number theory commands by typing
info(numlib).

ODEs and MuPAD to MATLAB Communication

Here is an illustrative example in which the communication between MATLAB and MuPAD is less than ideal and a specific problem is worked more easily in the latter than in the former.

Suppose that you want to consider solutions of the first-order ordinary differential equation

$$\frac{dy}{dt} + ty = 1.$$

This is a relatively simple first-order linear ordinary differential equation – albeit, the coefficients are not constant. To see the general nature of the solution curves, Figure 8.6 is a depiction of the vector field for the equation, which is achieved quite nicely in MATLAB using the commands **meshgrid** and **quiver** (see the online help for information on these commands):

```
>> [T, Y] = meshgrid(-4:0.2:4, -4:0.2:4);
>> S = 1 - T.*Y;
>> L = sqrt(1 + S.^2);
>> quiver(T, Y, 1./L, S./L, 0.5), axis equal tight
>> xlabel 't', ylabel 'y'
>> title 'Direction Field for dy/dt = 1 - ty'
```

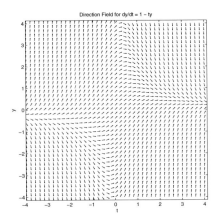

Figure 8.6. Vector Field for the ODE $y' + ty = 1$.

It's hard to be certain, but it appears that, proceeding from left to right, the solution curves are fairly flat, but then either increase or decrease sharply, after which they retreat back to their original position and then flatten out again.

MATLAB's symbolic differential equation solver **dsolve** can handle this equation. Since it produces a symbolic solution, we should be able to draw some actual solution curves using MATLAB's symbolic plotter **ezplot**. But, alas, the following code yields a slew of (red) error messages – only the first few lines of which are displayed:

```
>> syms c t y(t)
>> sol = dsolve(diff(y,t) == 1 - t*y, y(0) == c);
>> hold on
>> for cval = -2:0.5:2
   ezplot(subs(sol, c, cval), [-3 3])
end

Error using inlineeval (line 15)
Error in inline expression ==>
2.^(1./2).*pi.^(1./2).*exp(-t.^2./2).*erf(2.^(1./2).*t.*(i./2))
   .*(-i./2)- 2.*exp(-t.^2./2)
 Input must be real and full.
```

In order to get a handle on the problem, let's remove the semicolon on the second line of the code so that we can see the actual symbolic solution:

```
>> dsolve(diff(y,t) == 1 - t*y, y(0) == c)

ans =
c*exp(-t^2/2) - (2^(1/2)*pi^(1/2)*exp(-t^2/2)
   *erf((2^(1/2)*t*i)/2)*i)/2
```

Ah, there's the problem: although the solution function is real-valued, **dsolve**, which invokes MuPAD's symbolic solver, reports MuPAD's answer in terms of the special function *erf*, but with complex entries. In fact, the i's (one inside of *erf* and one outside of it) actually cancel; but **ezplot**, which is a MATLAB command, is unnerved by the presence of the complex parameter and reacts with error messages. We have spared the reader the full brunt of the error messages, but the lines displayed above clearly indicate **ezplot**'s inability to deal with complex parameters.

There are two methods that we could deploy in order to overcome this difficulty. One is to abandon **ezplot** and graph the solution curves numerically using **plot**. A slicker method, one that readers may prefer when MATLAB's symbolic solvers lead to difficulty, is to work directly in MuPAD. In fact, MuPAD's graphing routine, i.e., **plot**, can handle complex entries. What follows is a presentation of the two methods – and the resulting similar output: first, the MATLAB numerical solution (Figure 8.7), then the MuPAD symbolic solution (Figure 8.8):

```
>> syms c t y(t) yy(t, c) real
>> tt = sym(-3:0.01:3);
>> yy(t, c) = dsolve(diff(y,t) == 1 - t*y, y(0) == c);
>> figure; hold on
>> for d = -4:4
   plot(tt, eval(strrep(vectorize(yy(t,d/2)), 't', 'tt')))
end
```

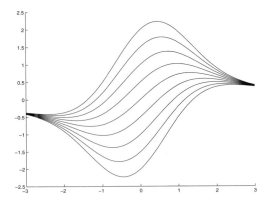

Figure 8.7. Numerical solution of the ODE $y' + ty = 1$ via MATLAB.

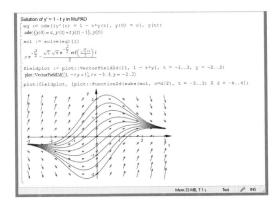

Figure 8.8. MuPAD Notebook containing a Symbolic Solution of the ODE $y' + ty = 1$.

MuPAD Programming

As with MATLAB, you can write small MuPAD programs to solve more complex problems. For example, here is an illustration of a special case of the Pollard $p - 1$ Factoring Method. In the following, p is a prime for which $p - 1$ has only small factors and q is a very large prime. The goal is to find factors of $n = pq$ without knowing *a priori* either p or q. This is a matter of some importance in cryptology, since factoring numbers such as n enables an attack on the RSA cryptosystem. Figure 8.9 shows a MuPAD Notebook that illustrates the method. The algorithm is encoded in the short loop. For more information on the algorithm, check your favorite number theory book or search online for "$p - 1$ factoring method." Note that the syntax for the **for** command is slightly different in MuPAD than in MATLAB. (It has to be followed by **do** and by **end_for**.)

The p-1 Factoring Method

We start with a prime p such that p-1 has only small factors.
This one was found by trial and error.

```
p := 200560490131
200560490131
ifactor(p-1)
2*3*5*7*11*13*17*19*23*29*31
```

Now choose another largish prime for q.

```
q := nextprime(153798454271389*10^10)
1537984542713890000000043
n := p*q
308458933700599683301628214101075633
b := 2
2
for j from 2 to 100 do
    b := powermod(b, j, n):
end_for
235344637899445984744233296525255523
gcd(b-1, n)
200560490131
```

Check:

```
ifactor(n)
200560490131*1537984542713890000000043
```

Figure 8.9. MuPAD Notebook illustrating the Pollard $p - 1$ factoring method.

Animations in MuPAD

It is quite easy to make animated graphics in MuPAD. If a MuPAD graphics command contains a "range" for a parameter and this range appears after the range(s) of values of the primary variable(s), then this parameter is treated as an animation parameter. Clicking on the graph in the notebook will activate the animation and cause the picture to move. You can start and stop the animation with the buttons that will "come to life" for this purpose in the notebook toolbar. For example, to produce a picture of a vibrating string, you can type:

```
plot(plot::Function2d(0.1*c*sin(x), x = 0..PI, c = 10..-10))
```

As in the case of movies produced with MATLAB, you can save as an AVI file, by clicking on **File:Export Graphics...** .

Chapter 9

Simulink

In this chapter, we describe Simulink, a MATLAB accessory for simulating dynamical processes and engineering systems. This brief introduction, together with the online documentation, should be enough to get you started using Simulink.

A Simple Differential Equation

If you want to learn about Simulink in depth, you can read the massive PDF document available at:

> www.mathworks.com/help/pdf_doc/simulink/sl_using.pdf

Here we give a brief introduction for the user who wants to get going with Simulink quickly. You start Simulink by clicking on the "Simulink Library" button on the MATLAB Desktop's Toolstrip, or simply by typing **simulink** in the Command Window. This opens the Simulink Library Browser, shown in Figure 9.1.

Figure 9.1. The Simulink Library Browser.

To begin to use Simulink, click **New:Model** from the **File** menu. This opens a blank model window. You create a Simulink model by copying units, called *blocks*, from the various Simulink libraries into the model window. We will explain how to use this procedure to model the homogeneous linear ordinary differential equation (ODE) $u'' + 2u' + 5u = 0$, which represents a damped harmonic oscillator.

First we have to figure out how to represent the equation in a way that Simulink can understand. One way to do this is as follows. Since the time variable is *continuous*, we start by opening the "Continuous" library by double-clicking on the icon at the top right of Figure 9.1. When opened, the "Continuous" library looks like Figure 9.2.

Figure 9.2. The Continuous Library.

Notice that u and u' are obtained from u' and u'' (respectively) by integrating. Therefore, drag two copies of the Integrator block into the model window, and line them up with the mouse. Simulink will automatically label the second one "Integrator1", to distinguish it from the first block called "Integrator". Note that each Integrator block has an input port and an output port. Align the output port of the Integrator with the input port of the Integrator1 and join them with an arrow, using the left button on the mouse. The arrow joining two blocks is called a *signal*. Double-click on this arrow and a little box will appear, in which you can type the label u'. (Alternatively, you can right-click on the arrow, select **Properties**, and type your label in the **Signal Name** box.) As before, add arrows leading into the Integrator (representing u'') and leading out of the Integrator1 (representing u); these will not connect to another block yet, so Simulink marks them with dotted lines to remind you that they are not fully operative. The idea so far is that u is obtained by integration from u', and u' is obtained by integration from u''. Your model window should now look like Figure 9.3.

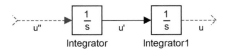

Figure 9.3. The First Stage of the Simulink Model.

Now we want to make use of the differential equation, which can be rewritten in the form $u'' = -5u - 2u'$. So we want to add other blocks to relate u'', the input to the first Integrator block, to u and u' according to this equation. For this purpose, we add two Gain blocks, which implement multiplication by a constant, and one Sum block, used for addition. These are all chosen from the "Commonly Used Blocks" library (upper left of Figure 9.1). Hooking them up the same way we did with the Integrator blocks gives a model window that looks something like Figure 9.4.

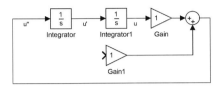

Figure 9.4. The Second Stage of the Simulink Model.

We need to go back and edit the properties of the Gain blocks, to change the constants by which they multiply from the default of 1 to -5 (in "Gain") and -2 (in "Gain1"). To do this, double-click on each Gain block in turn. A Block Parameters box, in which you can change the Gain parameter to whatever you need, will open. Next, we need to send u', the output of the first Integrator block, to the input port of block "Gain1". This presents a problem, since an Integrator block has only one output port and it's already connected to the next Integrator block. So we need to introduce a *branch line*. Position the mouse in the middle of the arrow connecting the two Integrators, hold down the CTRL key with one hand, simultaneously push down the left mouse button with the other hand, and drag the mouse around to the input port of the block entitled "Gain1". At this point, we're almost done; we just need a block for viewing the output. Open up the "Sinks" library and drag a copy of the Scope block into the model window. Hook this up with a branch line (again using the CTRL key) to the line connecting the second Integrator and the Gain block. Now you might want to relabel some of the blocks (by editing the text under each block), and also label some more of the arrows as before. The Simulink model as it appears in its model window is shown in Figure 9.5.

Now we're ready to run our simulation. First, it is a good idea to save the model, using **Save as...** from the **File** menu. We might choose to call it dampedosc. (MAT-LAB automatically adds the file extension .slx.) To see what is happening during the simulation, double-click on the Scope block to open an "oscilloscope" that will plot u as a function of t. Of course, we need to set initial conditions also; this can be done by double-clicking on the Integrator blocks and changing the line of the Block Parameters box that reads "Initial condition". For example, suppose that we set the initial condition for u' (in the first Integrator block) to 5 and the condition for u (in the second Integrator block) to 1. In other words, we are solving the system

$$u'' + 2u' + 5u = 0,$$
$$u(0) = 1,$$
$$u'(0) = 5,$$

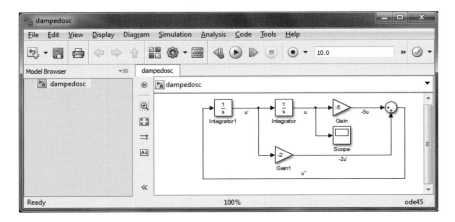

Figure 9.5. A Simulink Model for $u'' = -5u - 2u'$ in its Window.

which happens to have the exact solution

$$u(t) = 3e^{-t}\sin(2t) + e^{-t}\cos(2t).$$

Click on the green **Run** button that looks like ⊙ in the middle of the toolbar at the top of the model window. (You may first need to click in the model window itself to activate the model.) You should see in the Scope window something akin to Figure 9.6. This of course is simply the graph of the function $3e^{-t}\sin(2t) + e^{-t}\cos(2t)$. (By the way, you might need to change the scale on the vertical axis of the Scope window. Clicking on the **Autoscale** button 🔲 in this case gives the best results.) It is easy to go back and change some of the parameters and re-run the simulation again.

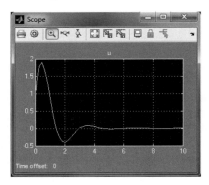

Figure 9.6. Scope Solution of $u'' = -5u - 2u'$, $u(0) = 1$, $u'(0) = 5$.

✓ Your first instinct might be to rely on the Derivative block, rather than the Integrator block, for simulating differential equations. But this has two drawbacks: it is harder to put in the initial conditions, and also numerical differentiation is much less stable than numerical integration.

Finally, suppose that you now want to study the *inhomogeneous* equation for "forced oscillations," $u'' + 2u' + 5u = g(t)$, where g is a specified "forcing" term. For this, all you have to do is add another block to the model from the Sources library. For example, if $g(t)$ is to represent "noise," drag the Band-Limited White Noise block from the "Sources" library into the model and drag it say to the top. Double-click on the Sum block to open up the Block Parameters box and change the "List of signs" to read $+ + +$. Your "Sum" block will now have three input ports instead of two so that you can hook up the extra source block. You should end up with a model as shown in Figure 9.7.

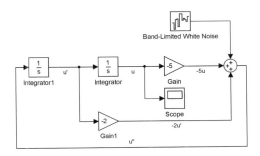

Figure 9.7. A Simulink Model of $u'' = -5u - 2u' + g(t)$.

The output from this revised model (with the default values of 0.1 for the noise power and 0.1 for the noise sample time) looks like Figure 9.8. The effect of noise on the system is clearly visible from the simulation.

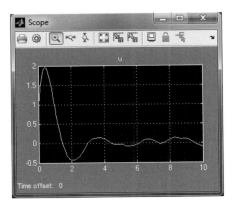

Figure 9.8. Scope Solution of an ODE with Noise.

☆ An Engineering Example

Now we present a more complicated example, typical of an engineering application of Simulink. Consider a crane used for loading shipping containers at a port. It can be schematically represented by Figure 9.9.

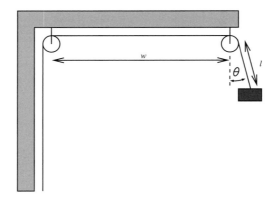

Figure 9.9. A Crane for Loading Containers.

Here we assume for simplicity that the radii of the pulleys and the size of the container are negligible, and that the pulley in the upper right can be moved horizontally, so that the crane operator can adjust the parameters w (the separation of the two pulleys) and l (the length of the cable holding the container) at will. We place our origin at the fixed pulley, so that, in terms of the parameters w, l, and θ (the angle between the cable and the vertical), the container is located at the point with coordinates

$$(x, y) = (w, 0) + (l\sin\theta, -l\cos\theta) = (w + l\sin\theta, -l\cos\theta). \qquad (9.1)$$

We want to simulate the result of lifting the container off the ground at point $(w_0, -l_0)$, dragging it, and then depositing it again at point $(w_1, -l_0)$. In particular, we want to control the side-to-side oscillations, so that the swinging container won't hit something or somebody. For simplicity, we assume that the only forces on the container are tension in the cable (which can vary with time) and gravity. (So we neglect wind and air resistance, for example.)

Let's make a further simplification: we assume that the crane operator moves the container in three steps, where in the first step the container is lifted straight up, in the second step l (the length of the cable connecting the container to the movable pulley) is held fixed, and in the third step the container is lowered straight down. Steps 1 and 3 are not especially interesting, so we concentrate on step 2, which means we assume that l is held constant and w is a function of time to be selected by the crane operator, with initial value w_0 and final value w_1. Let m be the mass of the container. To get the equations of motion, note that the kinetic energy K and gravitational potential energy

V of the container are given by

$$K = \frac{m}{2}((x')^2 + (y')^2) = \frac{m}{2}\left((w' + l(\cos\theta)\theta')^2 + (l(\sin\theta)\theta')^2\right)$$
$$= \frac{m}{2}\left((w')^2 + l^2(\theta')^2 + 2lw'(\cos\theta)\theta'\right) \tag{9.2}$$
$$V = mgy = -mgl\cos\theta.$$

In Lagrangian mechanics, the equation of motion is

$$\frac{\partial L}{\partial \theta} = \frac{d}{dt}\frac{\partial L}{\partial \theta'},$$

where the Lagrangian is $L = K - V$, so (after dividing out by the constant factor of m) we get

$$-gl\sin\theta - lw'(t)(\sin\theta)\theta' = \frac{d}{dt}\left(l^2\theta' + lw'(t)\cos\theta\right)$$
$$= l^2\theta'' + lw''(t)\cos\theta - lw'(t)(\sin\theta)\theta',$$

or (after dividing by l^2)

$$\theta'' + \frac{1}{l}w''(t)\cos\theta + \frac{g}{l}\sin\theta = 0, \tag{9.3}$$

which is the usual "pendulum equation" (see the discussion of the nonlinear pendulum in Chapter 11) with an extra term in it involving $w''(t)$, the horizontal acceleration of the crane. Now we can build our Simulink model for the crane. We want to experiment with various possibilities for $w(t)$ and see what effect these have on the measure of "swinging," $\theta(t)$. Let's use meters and seconds as our units and take $g = 9.81\,\text{m/sec}^2$, $l = 5\,\text{m}$, $w_0 = 0$, and $w_1 = 10\,\text{m}$. The time required to move the container should be on the order of a few minutes, or let's say about 200 sec. So we want w to be a continuous function, preferably with a continuous second derivative (since the force on the motor driving the pulley is proportional to w'' and should be a continuous function of time), with $w(0) = 0$ and $w(200) = 10$. A way to get such a function is to take

$$w(t) = 5 + a(t - 100) + b(t - 100)^3 + c(t - 100)^5.$$

This ensures that $w(100) = 5$ and that the graph of w has odd symmetry around the point $(100, 5)$. In fact, since only odd powers of $t - 100$ appear in the formula for $w(t) - 5$, it follows that $w(100 + t) - 5 = -(w(100 - t) - 5)$, and, on setting $t = 0$, we see that $2w(100) = 10$ or $w(100) = 5$. Putting $t = 100$ into this equation gives $w(200) - 5 = -w(0) + 5$, so we'll have $w(200) = 10$ if $w(0) = 0$. Thus, we solve for the coefficients a, b, and c so that $w(0) = w'(0) = w''(0) = 0$. Here is a solution obtained using MATLAB:

```
>> syms t a b c;
>> w = 5 + a*(t-100) + b*(t-100)^3 + c*(t-100)^5;
>> w0 = subs(w, t, 0);
>> w1 = subs(diff(w,t), t, 0);
>> w2 = subs(diff(w,t,2), t, 0);
```

```
>> [aa, bb, cc] = solve(w0, w1, w2);
>> w = subs(w, [a,b,c], [aa,bb,cc])

w =
(3*t)/32 - (t - 100)^3/160000 + (3*(t - 100)^5)/16000000000 - 35/8
```

A plot of this function is shown in Figure 9.10.

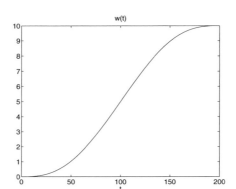

Figure 9.10. A Possible $w(t)$.

For this function, the formula for $w''(t)$ is:

```
>> diff(w, t, 2)

ans =
-3/80000*t+3/800+3/800000000*(t-100)^3
```

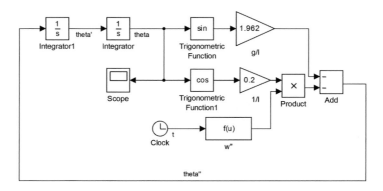

Figure 9.11. A Simulink Model for Studying the Motion of a Crane.

A possible Simulink model for studying this system is shown in Figure 9.11. Let's discuss how this model works and how to assemble it. We begin by rewriting equation

(9.3) in the form

$$\theta'' = -\frac{g}{l}\sin\theta - \frac{1}{l}w''(t)\cos\theta.$$

Thus we want to represent θ'' as a sum of two terms, each with a "minus sign" in front, which are assembled together in an Add block. (By the way, there is no difference between the Add and Sum blocks except for the shape of the representing icon.) The signal representing θ'' then enters the block called Integrator1 at the upper left, and the signal exiting the second Integrator represents θ itself. The Clock block outputs the value of t, which goes into a Function block (from the User-Defined Functions library) used for computing $w''(t)$. The Product and Trigonometric Function blocks, both from the Math Operations library, are self-explanatory. There are two Gain blocks, again from the Math Operations library, used to multiply $\sin\theta$ and $\cos\theta$ by the constants g/l and $1/l$, respectively. For convenience, we have renamed these two blocks to indicate which is which. To rename a block, click in the label under the block. You can then erase the old name and insert a new one. As in the example in the previous section, the Gain blocks have to be customized for the appropriate constants using the Block Parameters dialog box. Similarly, the Block Parameters dialog box for the "Trigonometric Function1" block can be used to switch from the sine function (the default) to the cosine function that we need. Since the crane starts at rest (with initial conditions $\theta(0) = \theta'(0) = 0$), we do not need to change the default initial condition of 0 in each of the Integrator blocks. We *do* need to customize the Function block by inserting the formula for $w''(u)$. (Simulink insists here that the independent variable be called u, not t.) Finally, we need to click on **Model Configuration Parameters** in the **Simulation** menu to change the maximum value of t from the default of 5 to the value 200 needed for this problem.

Running the simulation (and then clicking on the **Autoscale** button to rescale the graph) gives the Scope picture shown in Figure 9.12. If you look at the legend on the picture, you will see that the maximum value of θ is on the order of 2×10^{-4} radians, which is pretty small, even when multiplied by the length of the cable, $l = 5\,\text{m}$. So we conclude that the shaking of the container is not a serious problem in this case.

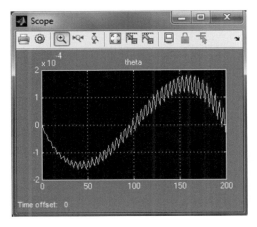

Figure 9.12. Scope Output of the Oscillations of a Crane.

☆ Communication with the Workspace

The examples we have discussed should suffice to give you an idea of how to use Simulink as a stand-alone simulation tool. But part of the power of Simulink comes from the way it can be combined with MATLAB. For example, you can run a Simulink model from within an M-file, or do further processing within MATLAB on the output of a Simulink simulation. In this section, we discuss some of the key commands for combining Simulink with MATLAB.

We'll begin with a practical example. Suppose that we want to recompute the formula for $w(t)$ in the crane example so that w reaches the value of 10 at $t = 20$ instead of at $t = 200$. In other words, we want to retain the shape of $w(t)$ in Figure 9.10, but compress the scale on the horizontal axis by a factor of 10. What happens to the oscillations now that the crane moves the container more rapidly? Is the process now dangerous? To answer this question, we replace the old $w(t)$ by

```
>> w = simplify(subs(w, t, 10*t))

w =
(t^3*(3*t^2 - 150*t + 2000))/160000
```

and replace the old $w''(t)$ by:

```
>> simplify(diff(w, t, 2))

ans =
(3*t*(t - 10)*(t - 20))/8000
```

The first thing we need to do is to make a change in the parameters of the block labeled f(u) in Figure 9.11. It is possible to do this with the commands **get_param** and **set_param** from the command line or from an M-file, not just by bringing up the Block Parameters box. This is especially convenient if you need to a run a simulation many times with various values of the parameters. Sometimes it helps to use the commands **gcs** ("get current system") and **find_system** to locate the names of the relevant blocks. Here's an example, based on the supposition that we've already opened the model in Figure 9.11:

```
>> gcs

ans =
crane

>> find_system(gcs, 'Type', 'block')

ans =
    'crane/1//1'
    'crane/Add'
    'crane/Clock'
    'crane/Integrator'
    'crane/Integrator1'
    'crane/Product'
    'crane/Scope'
    [1x28 char]
    [1x29 char]
    'crane/g//1'
```

```
'crane/w"'
```

This gives us the names of all of the blocks (the ones that are identified as $1 \times$ 28 and 1×29 character arrays are `'crane/Trigonometric Function'` and `'crane/Trigonometric Function1'`). In this case, we need to change the parameters of the block entitled `'crane/w"'`. (If we wanted to change the cable length, we would also have to change both `'crane/1//1'` and `'crane/g//1'`.) To see the current value of the `'Expr'` parameter of this block, which encodes the function $w''(t)$, we type:

```
>> get_param('crane/w"', 'Expr')
```

```
ans =
-3/80000*u+3/800+3/800000000*(u-100)^3
```

So we can reset this with:

```
>> set_param('crane/w"', 'Expr', '3*u*(u - 10)*(u - 20)/8000')
```

⇨   **Note that most parameters of most Simulink blocks need to be specified as strings, not as numbers or symbolic expressions.**

To rerun the model, instead of using the **Run** button, we can use the command **sim** from the command line or within an M-file. The simplest form of this command just runs the model with the existing parameters. But we can also use this command to change the stopping or starting time and to store values of specified signals in a form that can be accessed by the workspace. For example, in our situation, we could type:

```
>> craneout=sim('crane','SaveTime','on','TimeSaveName','t', ...
'SaveState','on','StateSaveName','theta','StopTime','20');
t = craneout.find('t'); theta = craneout.find('theta');
```

We could then type

```
>> plot(t, theta)
```

to plot the results, giving the picture in Figure 9.13.

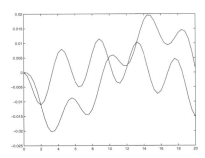

Figure 9.13. Plot of $\theta(t)$ and $\theta'(t)$ for the Revised Crane Model.

Here the curve that ends up on top represents $\theta(t)$, and the other curve represents $\theta'(t)$. Or to duplicate what would be seen in the Scope window, we can replace **plot** by **simplot**, getting Figure 9.14:

```
>> simplot(t, theta(:,1))
```

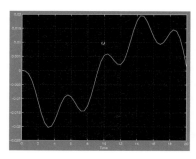

Figure 9.14. "Scope" Plot of $\theta(t)$ for the Revised Crane Model.

✓ Incidentally, if you draw a plot with **simplot**, it always appears in the figure window as in Figure 9.14, with a light-colored plot on a black background. But if you print the plot (either to the printer or to a file), it will sometimes (depending on your **print** defaults) appear with the colors reversed, i.e., as a dark plot on a white background. To undo this color reversal, you need to insert the command

```
>> set(gcf, 'InvertHardcopy', 'off')
```

before executing the **print** command.

The result of our analysis shows that if the shipping container is moved in 20 seconds instead of 200, the oscillations are on the order of 0.02 radians. Thus the container will swing back and forth about 0.10 m or 10 cm, which is again a reasonable amount. We leave it to the reader to see what happens if the container is moved in only 2 seconds!

Chapter 10

☆ GUIs

With MATLAB, you can create your own *Graphical User Interface*, or GUI, which consists of a Figure window containing menus, buttons, text, graphics, etc., that a user can manipulate interactively with the mouse and keyboard. There are two main steps in creating a GUI: one is designing its layout, and the other is writing *callback functions* that perform the desired operations when the user selects different features.

GUI Layout and GUIDE

Specifying the location and properties of various objects in a GUI can be done with commands such as **uicontrol**, **uimenu**, and **uicontextmenu** in an M-file. MATLAB also provides an interactive tool (a GUI itself!) called GUIDE (this stands for Graphical User Interface Development Environment) that greatly simplifies the task of building a GUI. We will describe here how to get started writing GUIs with the MATLAB 8 version of GUIDE, which is similar to the version in MATLAB 7 but has some new features.

✓   GUIDE saves your GUI as two files: the layout is stored in a binary `.fig` file, and the callback functions are stored in an M-file. The resulting GUI may not be compatible with significantly older versions of MATLAB. If your goal is to create a robust GUI that many different users can use with different versions of MATLAB, you may be better off in the long run writing the GUI from scratch as an M-file.

To open GUIDE, select **New:Graphical User Interface** from the FILE section of the HOME tab in the MATLAB Desktop, or type **guide** in the Command Window. You will see the GUIDE Quick Start dialog box, shown in Figure 10.1.

Note that there are two tabs at the top. The left-hand one, "Create New GUI", is open by default. You can start by selecting one of the provided templates. We recommend also selecting "Save new figure as:" and giving your GUI a name right away. Then, clicking "OK" will open both a `.fig` file in the Layout Editor and a corresponding M-file in the Editor/Debugger. For purposes of the example here, we'll assume that you select the third template, "GUI with Axes and Menu" and enter `plotgui.fig` for the file name. The Layout Editor that opens looks like Figure 10.2. Let's say that you want to design a GUI that will accept a MATLAB plotting command as input, display the corresponding output, and have buttons for making various changes in the appearance of the output.

The buttons at the left of the Layout Editor are used for inserting various kinds of objects. You can add features to your GUI by clicking on one of these buttons, then

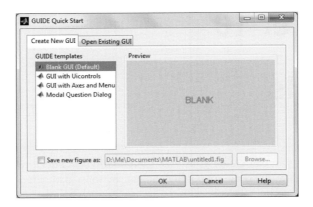

Figure 10.1. The GUIDE Quick Start Dialog Box.

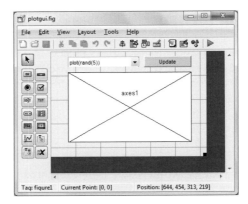

Figure 10.2. The Layout Editor.

moving to a desired location in the grid, and clicking again to place an object on the grid. (Move the mouse pointer over the buttons to see what type of object each button creates.) Once you have placed an object on the grid, you can click and drag on the middle of the object to move it, or click and drag on a corner to resize the object. After you have placed several objects, you can select multiple objects by clicking and dragging on the background grid to enclose them with a rectangle. Then you can move the objects as a block with the mouse, or align them by selecting **Align Objects** from the toolbar or **Tools** menu. To break the block back into separate objects, click on the background grid again.

✓   The lower right button labeled with an "X" appears only in the Windows version of MATLAB. It creates an "ActiveX" control, which would be functional only for other Windows users.

To change properties of an object such as its color, the text within it, etc., you must open the *Property Inspector* window. To do so, you can double-click on an object, or choose **Property Inspector** from the toolbar or the **View** menu. Then click on the object you want to alter. You can leave the Property Inspector open throughout your GUIDE session and go back and forth between it and the Layout Editor.

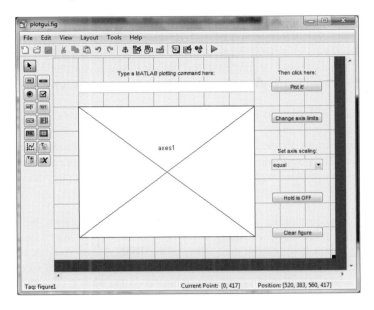

Figure 10.3. The Layout Editor with Design for a "Plot GUI".

Figure 10.3 shows what the Layout Editor window looks like after several objects have been placed and their properties adjusted. Let us describe how we created the objects that make up the GUI.

The two boxes on the top – as well as the one labeled "Set axis scaling:" – are *Static Text* boxes, which the user of the GUI will not be allowed to manipulate. To create each of them, we first clicked on the "Static Text" button – the one to the left of the grid labeled "TXT" – and then clicked in the grid where we wanted to add the text. Next, to set the text for the box we opened the Property Inspector and clicked on the square button next to "String", which opens a new window that contains editable default text. (Another option is to click in the text to the right of the square button and edit it there.) Finally, we resized each box according to the length of its text (by clicking on a corner and dragging it).

The buttons labeled "Plot it!", "Change axis limits", and "Clear figure" are all *Push Button* objects, created using the button labeled "OK". To make these buttons all the same size, we first created one of them and then, after sizing it, we duplicated it (twice) by clicking the right mouse button on the existing object and selecting **Duplicate**. We then moved each new Push Button to a different position and changed its text in the same way as we did for the Static Text boxes.

The blank box near the top of the grid is an *Edit Text* box, which allows the user to enter text. We created it with the button labeled "EDIT" and then cleared its default text in the same way that we changed text before. Below the Edit Text box is a large

Axes box, created with the button labeled with a graph, and in the lower right the button labeled "Hold is OFF" is a *Toggle Button*, normally created with the button labeled "TGL". In order to make it the same size as the *Push Button* objects, we actually duplicated another *Push Button* and then changed its "Style" in the Property Inspector from "pushbutton" to "togglebutton". For toggling (on-off) commands, you could also use a *Radio Button* or a *Checkbox*, which are denoted, respectively, by the buttons with a dot and a check mark in them. Finally, the box on the right that says "equal" is a *Popup Menu* – its button in the Layout Editor is right below the *Edit Text* button. Popup Menus and *Listbox* objects allow you to let the user choose among several options.

We moved, resized, and in most cases changed the properties of each object in the fashion described above. In the case of the Popup Menu, after we selected the "String" button in the Property Inspector, we entered into the window that appeared three words on three separate lines: **equal**, **normal**, and **square**. Using multiple lines is necessary in order to give the user multiple choices in a Popup Menu or Listbox object.

✓   In addition to populating your GUI with the objects we described above, you can create a menu bar for it using the Menu Editor, which you can open by selecting **Menu Editor** from the toolbar or the **Tools** menu. You can also use the Menu Editor to create a *context menu* for an object; this is a menu that appears when you click the right mouse button on the object. See the online documentation for GUIDE to learn how to use the Menu Editor.

We also gave our GUI a title, which will appear in the title bar of its window, as follows. We clicked on the background grid in the Layout Editor to select the entire GUI (as opposed to an object within it) and went to the Property Inspector. There we changed the text to the right of "Name" from "plotgui" to "Simple Plot GUI".

Saving and Running a GUI

To save a GUI, as usual type CTRL+S or select **Save** from the **File** menu. If you did not choose a name for your GUI before, type one now without any extension. The corresponding .fig file and M-file are both saved. The M-file contains templates for the callback functions that will make the objects we created functional; we describe how to add functionality to the M-file in the next section.

⇨   **The instructions in this and the following section assume the default settings of the GUI Options..., which you can access from the Tools menu. We assume in particular that "Generate FIG file and MATLAB file" and "Generate callback function prototypes" are selected.**

Once it has been saved, you can run the GUI from the Command Window by typing its name, in our case **plotgui**, whether or not GUIDE is running. Both the .fig file and the .m file must be in your current folder or search path. You can also run the GUI from the Layout Editor by typing CTRL+T or selecting **Run** from the **Tools** menu. A copy of the GUI will appear in a separate window, without all the surrounding menus and buttons of the Layout Editor. (If you have added new objects since the last time you saved or activated the GUI, the M-file associated with the GUI will also be brought to the front.) Figure 10.4 shows how the GUI we created above looks when activated.

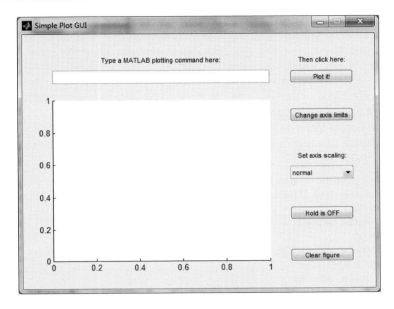

Figure 10.4. The "Simple Plot GUI" Activated.

Notice that the appearance of the GUI differs slightly from that in the GUIDE window; in particular, the background grid is absent. When creating your own GUI, you may find that an object doesn't appear the way you expected in the activated GUI; if so, go back to the GUIDE window and make adjustments. The changes you make will not immediately appear in the active GUI; to see their effect, you must activate the GUI again.

The objects you create in the Layout Editor are inert within that window – you can't type text in the Edit Text box, you can't see the additional options by clicking on the Popup Menu, etc. But, in an activated GUI window, objects such as Toggle Buttons and Popup Menus will respond to mouse clicks. However, they will not actually perform any operations until you write a callback function for each of them.

GUI Callback Functions

When you are ready to create a callback function for a given object, make sure you've saved your GUI. Then look at the Property Inspector for the object in the Layout Editor, and under Tag you will see the tag by which it is identified in the M-file associated with the GUI. You can search for the corresponding tag in the M-file, or right-click on the object in the Layout Editor and select **View Callbacks:Callback**.

You will see a small block of text that looks like this:

```
% --- Executes on button press in pushbutton1.
function pushbutton1_Callback(hObject, eventdata, handles)
% hObject      handle to pushbutton2 (see GCBO)
% eventdata    reserved - to be defined in a future version of MATLAB
% handles      structure with handles and user data (see GUIDATA)
```

All you need to do now to bring this Push Button to life is to add the commands that you want performed when the user clicks on the button. Of course, you also need to save the M-file, which you can do in the usual way from the Editor/Debugger. Each time you save or activate a GUI, a block of five lines like the ones above is automatically added to the GUI's M-file for any new objects or menu items that you have added to the GUI and that should have callback functions.

For the Popup Menu on the right-hand side of the GUI, we put the following lines into its callback function (modified from the template that came with the sample GUI of Figure 10.2):

```
popup_sel_index = get(handles.popupmenu1, 'Value');
switch popup_sel_index
    case 1
        axis equal
    case 2
        axis normal
    case 3
        axis square
end
```

As we described in Chapter 5, you can use **get** to retrieve the current setting of a property of a graphics object. When you use the callback templates provided by GUIDE as we have described, the variable **hObject** will contain the handle (the required first argument of **get** and **set**) for the object associated with the callback, and **handles** will contain handles to all the objects in the GUI. Thus, in the callback function above, we could also use **hObject** instead of **handles.popupmenu1**. (If you are using another method to write callback functions, you can use the MATLAB command **gcbo** in place of **hObject**.) Each time the user of a GUI selects an item from a Popup Menu, MATLAB sets the "Value" property of the object to the line number selected and runs the associated callback function. For our sample GUI, line 1 of the Popup Menu says "equal", and, if the user selects line 1, the callback function above runs **axis equal**; line 2 says "normal", etc.

✓   You may have noticed that the Popup Menu in Figure 10.4 says "normal" rather than "equal" as in Figure 10.3; that's because we set its "Value" property to 2 when we created the GUI, using the Property Inspector. In this way, you can make the default selection something other than the first item in a Popup Menu or Listbox.

For the Push Button labeled "Plot it!", we wrote the following callback function:

```
axes(handles.axes1);
eval(get(handles.edit1, 'String'))
```

Here **handles.axes1** and **handles.edit1** are the handles for the Axes box and for the Edit Text box, respectively. Again, these variables are provided by the callback templates in GUIDE, and if you do not use this feature, you can generate the appropriate handles with **findobj(gcbf, 'Tag', 'axes1')** and **findobj(gcbf, 'Tag', 'edit1')**, respectively. The first line of the callback function above is copied from the template in the sample GUI of Figure 10.2, and directs plotting to the Axes box. (We chose, however, not to include the command **cla** from the template, because it would counteract the effect of **hold on** when we later set up the Toggle Button.) The second line uses **get** to find the text in the Edit Text box and then runs the corresponding command with **eval**.

✓   We could also associate a callback function with the Edit Text box; this function would be run each time the user presses the ENTER key after typing text in the box. The callback function **eval(get(hObject, 'String'))** would run the command just typed, providing an alternative to (or making superfluous) the "Plot it!" button.

For the Push Button labeled "Change axis limits...", we used the callback function below to open a dialog box and let the user type in new axis limits:

```
prompt = {'Enter horizontal range:', 'Enter vertical range:'};
current = axis;
default = {num2str(current(1:2)), num2str(current(3:4))};
answer = inputdlg(prompt, 'Axis Limits', 1, default);
if size(answer)
    axis([str2num(answer{1}) str2num(answer{2})])
end
```

See the online help for **inputdlg** to understand how we constructed these commands, in particular the use of cell arrays. The resulting dialog box appears in Figure 10.5. The existing horizontal and vertical ranges are extracted from **axis** and displayed, and the user has an opportunity to change one or both. If the user clicks "OK", then the new ranges are set with **axis**.

Figure 10.5. A Dialog Box.

✓   MATLAB has a number of dialog boxes that you can use either as callback functions or in an ordinary M-file. For example, you can use **inputdlg** in place of **input**. See the online help for **uitools** for a list of available dialog boxes.

We used the following callback function for the Toggle Button labeled "Hold is OFF":

```
if get(hObject, 'Value')
    hold on
    set(hObject, 'String', 'Hold is ON');
else
    hold off
    set(hObject, 'String', 'Hold is OFF');
end
```

We get the "Value" property of the Toggle Button in the same way as in the Popup Menu callback function above, but for a Toggle Button this value is either 0 if the button is "out" (the default) or 1 if the button is pressed "in." (Radio Buttons and Checkboxes also have a "Value" property of either 0 or 1.) When the user first presses the Toggle Button, the value is set to 1, so the callback function above runs **hold on** and resets the string displayed on the Toggle Button to reflect the change. The next time the user presses the button, these operations are reversed.

Here is our callback function for the Push Button labeled "Clear figure":

```
set(handles.edit1, 'String', '');
set(handles.togglebutton1, 'String', 'Hold is OFF', 'Value',0);
set(handles.popupmenu1, 'Value', 2);
cla reset
```

These commands clear the text in the Edit Text box, reset the Toggle Button and the Popup Menu to their default states, and clear the Axes box.

Finally, with an object in a GUI you can also associate a variant of a callback function called a "ButtonDownFcn" instead. Such a function will begin something like

```
% --- Executes on mouse press over axes background.
function axes1_ButtonDownFcn(hObject, eventdata, handles)
```

and will be run when the user clicks in the axes window or, in the case of other objects such as a Push Button, clicks with the right mouse button (as opposed to the left mouse button for the callback function). To create such a function, you can right-click on the object in the Layout Editor and select **View Callbacks:ButtonDownFcn**. You can associate functions with several other types of user events as well; to learn more, see the online documentation, or experiment by clicking the right mouse button on various objects and on the grid behind them in the Layout Editor.

Chapter 11

Applications

In this chapter, we present examples showing you how to apply MATLAB to problems in several disciplines. Each example is presented in the form of a MATLAB M-file, published to LaTeX. We modified the default style sheet for publishing to LaTeX to adjust the spacing between input, output, and text. We also made a few minor adjustments to the published LaTeX code to improve line breaks. These examples are illustrations of the kinds of polished, integrated documents that you can create with MATLAB. The examples are:

- Mortgage Payments

- Monte Carlo Simulation

- Population Dynamics

- Linear Programming

- Image Compression and Manipulation

- Processing and Analysis of Financial Data

- The 360° Pendulum

- ☆ Numerical Solution of the Heat Equation

- ☆ A Model of Traffic Flow

We have not explained all the MATLAB commands that we use; you can learn about the new commands from the online help. Simulink is used in *A Model of Traffic Flow* and as an optional accessory in *Population Dynamics* and *Numerical Solution of the Heat Equation*. The example on *Linear Programming* also requires an M-file found (in slightly different forms) in the Optimization and Simulink toolboxes. The examples require different levels of mathematical background and expertise in other subjects. *Mortgage Payments* and *Population Dynamics* use only high-school mathematics. *Monte Carlo Simulation* and *Processing and Analysis of Financial Data* use some probability and statistics and use the Statistics Toolbox (though there are ways of doing the calculations without it); *Linear Programming* and *Image Compression and Manipulation*, some linear algebra (the latter also includes some commands from the Image Processing Toolbox); *The 360° Pendulum*, some ordinary differential equations; *Numerical Solution of the Heat Equation*, some partial differential equations; and *A Model of Traffic Flow*, differential equations, linear algebra, and familiarity with the function e^z for z a complex number. There is also a little Fourier analysis in

141

Processing and Analysis of Financial Data. Even if you don't have the background for a particular example, you should be able to learn something about MATLAB from the presentation.

Mortgage Payments

We want to understand the relationships between the mortgage payment of a fixed rate mortgage, the principal (the amount borrowed), the annual interest rate, and the period of the loan. We are going to assume (as is usually the case in the United States) that payments are made monthly, even though the interest rate is given as an annual rate. Let's define

```
peryear = 1/12; percent = 1/100;
```
So the number of payments on a 30-year loan is

```
30*12
```

```
ans =
```

```
   360
```

and an annual percentage rate of, say, 8% comes out to a monthly rate of:

```
8*percent*peryear
```

```
ans =
```

```
   0.0067
```

Now consider what happens with each monthly payment. Some of the payment is applied to interest on the outstanding principal amount, P, and some of the payment is applied to reduce the principal owed. The total amount, R, of the monthly payment, remains constant over the life of the loan. So if J denotes the monthly interest rate, then the amount applied to reduce the principal is $R - JP$, and the new principal after the payment is applied is

$$P - R + JP = P(1 + J) - R = Pm - R,$$

where $m = 1 + J$. So a table of the amount of the principal still outstanding after n payments is tabulated as follows for a loan of initial amount A, for n from 0 to 6:

```
syms m J P R A n x y
disp('No. of Pays  Remaining Principal');
P = A;
for n = 0:6
    disp([num2str(n), '                    ', char(P)])
    P = simplify(-R + P*m);
end
```

```
No. of Pays  Remaining Principal
0            A
1            A*m  -  R
2              - R - m*(R - A*m)
3              - R - m*(R + m*(R - A*m))
4              - R - m*(R + m*(R + m*(R - A*m)))
5              - R - m*(R + m*(R + m*(R + m*(R - A*m))))
6              - R - m*(R + m*(R + m*(R + m*(R + m*(R - A*m)))))
```

We can write this in a simpler way by noticing that

$$P = Am^n + (\text{terms divisible by } R)$$

For example, with $n = 7$ we have:

`factor(P - A*m^7)`

`ans =`

`-R*(m^6 + m^5 + m^4 + m^3 + m^2 + m + 1)`

But, on the other hand, the quantity inside the parentheses is the sum of a geometric series

$$\sum_{k=1}^{n-1} m^k = \frac{m^n - 1}{m - 1}.$$

So we see that the principal after n payments can be written as

$$P = Am^n - R(m^n - 1)/(m - 1).$$

Now we can solve for the monthly payment amount R under the assumption that the loan is paid off in N installments, i.e., P is reduced to 0 after N payments:

```
syms N;
solve(A*m^N - R*(m^N - 1)/(m - 1), R)
R = subs(ans, m, J + 1)
```

`ans =`

`(A*m^N*(m - 1))/(m^N - 1)`

`R =`

`(A*J*(J + 1)^N)/((J + 1)^N - 1)`

For example, with an initial loan amount $A = \$150,\!000$ and a loan lifetime of 30 years (360 payments), we get the following table of payment amounts as a function of annual interest rate:

```
format bank;
disp('      Rate (%)      Monthly Payment ($)');
for rate = 1:10,
  disp([rate,double(subs(R, {A, N, J}, ...
      {150000, 360, rate*percent*peryear}))])
end
```

Rate (%)	Monthly Payment ($)
1.00	482.46
2.00	554.43
3.00	632.41
4.00	716.12
5.00	805.23
6.00	899.33
7.00	997.95
8.00	1100.65
9.00	1206.93
10.00	1316.36

Note the use of **format bank** to write the floating-point numbers with two digits after the decimal point.

There's another way to understand these calculations that's a little slicker, and that uses MATLAB's linear-algebra capability. Namely, we can write the fundamental equation

$$P_{\text{new}} = P_{\text{old}} m - R$$

in matrix form as

$$v_{\text{new}} = B v_{\text{old}},$$

where

$$v = \begin{pmatrix} P \\ 1 \end{pmatrix}$$

and

$$B = \begin{pmatrix} m & -R \\ 0 & 1 \end{pmatrix}.$$

We can check this using matrix multiplication:

```
syms R P; B = [m -R; 0 1]; v = [P; 1]; B*v
```

```
ans =

   P*m - R
       1
```

which agrees with the formula we had above. Thus, the column vector **[P; 1]** resulting after n payments can be computed by left-multiplying the starting vector **[A; 1]** by the matrix B^n. Assuming that $m > 1$, that is, that the rate of interest is positive, the calculation

```
[eigenvectors, diagonalform] = eig(B)
```

```
eigenvectors =

[ R/(m - 1), 1]
[         1, 0]
```

```
diagonalform =

[ 1, 0]
[ 0, m]
```

shows us that the matrix B has eigenvalues m, 1, and corresponding eigenvectors

$$\begin{pmatrix} 1 \\ 0 \end{pmatrix} \text{ and } \begin{pmatrix} 1 \\ (m-1)/R \end{pmatrix} = \begin{pmatrix} 1 \\ J/R \end{pmatrix}.$$

Now we can write the vector **[A; 1]** as a linear combination of the eigenvectors:

$$\begin{pmatrix} A \\ 1 \end{pmatrix} = x \begin{pmatrix} 1 \\ 0 \end{pmatrix} + y \begin{pmatrix} 1 \\ J/R \end{pmatrix}.$$

We can solve for the coefficients:

```
[x, y] = solve(A == x*1 + y*1, 1 == x*0 + y*J/R, x, y)

x =

-(R - A*J)/J


y =

R/J
```

and so

$$\begin{pmatrix} A \\ 1 \end{pmatrix} = (A - (R/J)) \begin{pmatrix} 1 \\ 0 \end{pmatrix} + (R/J) \begin{pmatrix} 1 \\ J/R \end{pmatrix}$$

and

$$B^n \begin{pmatrix} A \\ 1 \end{pmatrix} = (A - (R/J))m^n \begin{pmatrix} 1 \\ 0 \end{pmatrix} + (R/J) \begin{pmatrix} 1 \\ J/R \end{pmatrix}.$$

So the principal remaining after n payments is:

$$P = ((AJ - R)m^n + R)/J = Am^n - R(m^n - 1)/J.$$

This is the same result that we obtained earlier. To conclude, let's determine the amount of money A you can afford to borrow as a function of what you can afford to pay as the monthly payment R. We simply solve for A in the equation that $P = 0$ after N payments:

```
solve(A*m^N - R*(m^N - 1)/(m - 1), A)

ans =

(R*(m^N - 1))/(m^N*(m - 1))
```

For example, if we were shopping for a house and could afford to pay $1500 per month for a 30-year fixed-rate mortgage, the maximal loan amount as a function of the interest rate is given by:

```
disp('      Rate (%)      Maximal Loan ($)');
for rate = 1:10,
  disp([rate,double(subs(ans, {R, N, m}, ...
      {1500, 360, 1 + rate*percent*peryear}))])
end
```

Rate (%)	Maximal Loan ($)
1.00	466360.60
2.00	405822.77
3.00	355784.07
4.00	314191.86
5.00	279422.43
6.00	250187.42
7.00	225461.35
8.00	204425.24
9.00	186422.80
10.00	170926.23

Monte Carlo Simulation

In order to make statistical predictions about the long-term results of a random process, it is often useful to do a simulation based on the underlying probabilities. This procedure is referred to as the Monte Carlo method. We will consider two examples here – one of which relies on a uniform probability distribution, the other on a normal or Gaussian probability distribution.

Casino Game

For the first example, consider a casino game in which a player bets against the casino, and the casino wins 51% of the time. The question is as follows: how many games have to be played before the casino is reasonably sure of coming out ahead? This scenario is common enough that mathematicians long ago figured out very precisely what the statistics are, but here we want to illustrate how to get a good idea of what can happen in practice without having to absorb a lot of mathematics.

First we construct an expression that computes the net revenue to the casino for a single game, based on a random number chosen between 0 and 1 by the MATLAB function **rand**. If the random number is less than or equal to 0.51, the casino wins one betting unit, whereas if the number exceeds 0.51, the casino loses one unit. (In a high-stakes game, each bet may be worth $1000 or more. Thus it is important for the casino to know how bad a losing streak it may have to weather in order to turn a profit so that it doesn't go bankrupt first!) Here is an expression that returns 1 if the output of **rand** is less than 0.51 and −1 if the output of **rand** is greater than 0.51 (it will also return 0 if the output of **rand** is exactly 0.51, but this is extremely unlikely).

```
revenue = sign(0.51 - rand)

revenue =

    -1
```

To simulate several games at once, say ten games, we can generate a vector of ten random numbers with the command **rand(1,10)** and then apply the same operation:

```
revenues = sign(0.51 - rand(1, 10))

revenues =

    1    -1     1    -1    -1     1     1    -1     1    -1
```

Each 1 represents a game that the casino won, and each -1 represents a game that it lost. For a larger number of games, say 100, we can let MATLAB sum the revenue from the individual bets as follows:

```
profit = sum(sign(0.51 - rand(1, 100)))

profit =

    -4
```

The output represents the net profit (or loss, if negative) for the casino after 100 games. On average, every 100 games the casino should win 51 times and the player(s) should win 49 times, so the casino should make a profit of 2 units (on average). Let's see what happens in a few trial runs:

```
profits = sum(sign(0.51 - rand(100, 10)))

profits =

    14   -12     6     2    -4     0   -10    12     0    12
```

We see that the net profit can fluctuate significantly from one set of 100 games to the next, and there is a sizable probability that the casino has lost money after 100 games. To get an idea of how the net profit is likely to be distributed in general, we can repeat the experiment a large number of times and make a histogram of the results. The following function computes the net profits for k different trials of n games each:

```
profits = @(n,k) sum(sign(0.51 - rand(n, k)))

profits =

    @(n,k) sum(sign(0.51 - rand(n, k)))
```

What this function does is to generate an $n \times k$ matrix of random numbers and then perform the same operations as above on each entry of the matrix to obtain a matrix with entries 1 for bets the casino won and -1 for bets it lost. Finally, it sums the columns of the matrix to obtain a row vector of k elements, each of which represents the total profit from a column of **n** bets.

Now we make a histogram of the output of **profits** using $n = 100$ and $k = 100$. Theoretically, the casino could win or lose up to 100 units, but in practice we find that the outcomes are almost always within 30 or so of 0. Thus, we let the bins of the histogram range from -40 to 40 in increments of 2 (since the net profit is always even after 100 bets):

hist(profits(100, 100), -40:2:40); axis tight

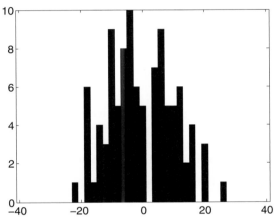

The histogram confirms our impression that there is a wide variation in the outcomes after 100 games. It looks like the casino is about as likely to have lost money as it is to have profited. However, the distribution shown above is irregular enough to indicate that we really should run more trials to see a better approximation to the actual distribution. Let's try 1000 trials:

hist(profits(100, 1000), -40:2:40); axis tight

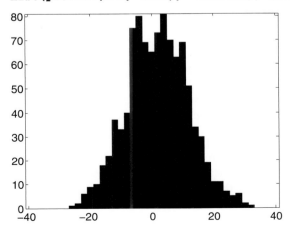

According to the "Central Limit Theorem," when both n and k are large, the histogram should be shaped like a "bell curve," and we begin to see this shape emerging above. Let's move on to 10,000 trials:

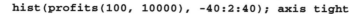

`hist(profits(100, 10000), -40:2:40); axis tight`

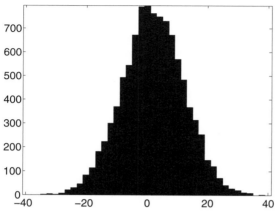

Here we see very clearly the shape of a bell curve. Though we haven't gained that much in terms of knowing how likely the casino is to be behind after 100 games, and how large its net loss is likely to be in that case, we do gain confidence that our results after 1000 trials are a good depiction of the distribution of possible outcomes.

Now we consider the net profit after 1000 games. We expect on average the casino to win 510 games and the player(s) to win 490, yielding a net profit of 20 units. Let's start with 1000 trials:

`hist(profits(1000, 1000), -100:10:150); axis tight`

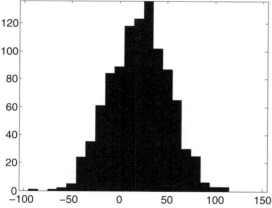

Though the range of values we observe for the profit after 1000 games is larger than the range for 100 games, the range of possible values is ten times as large, so that relatively speaking the outcomes are closer together than before. This reflects the theoretical principle (also a consequence of the Central Limit Theorem) that the average "spread" of outcomes after a large number of trials should be proportional to the square root of the number n of games played in each trial. This is important for the casino, since if the spread were proportional to n, then the casino could never be too sure of making a profit. When we increase n by a factor of 10, the spread should only increase by a factor of $\sqrt{10}$, or a little more than 3.

Notice that, after 1000 games, the casino is definitely more likely to be ahead than behind. However, the chances of being behind still look sizable. Let's repeat the simulation with 10,000 trials to be more certain of our results:

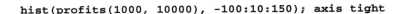

```
hist(profits(1000, 10000), -100:10:150); axis tight
```

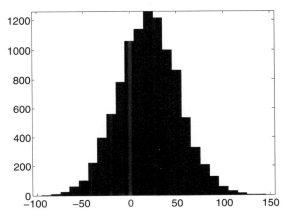

We see the bell-curve shape emerging again. Though it is unlikely, the chance that the casino is behind by more than 50 units after 1000 games is not insignificant. If each unit is worth $1000, then we might advise the casino to have at least $100,000 cash on hand in order to be prepared for this possibility. Maybe even that is not enough – to see we have to experiment further.

Let's see now what happens after 10,000 games. We expect on average the casino to be ahead by 200 units at this point, and, on the basis of our earlier discussion, the range of values we use to make the histogram needs to go up only by a factor of three or so from the previous case. Let's go straight to 10,000 trials. We might be tempted to run **profits(10000, 10000)**, but notice that this involves computing an array of 100 million numbers. While most computers can now store this many numbers in memory, using this much memory can slow MATLAB down. It may be more efficient to perform the 10,000 trials in batches of 1000, using a loop to assemble the results into a single vector:

```
profitvec = [];
for j = 1:10
    profitvec = [profitvec profits(10000, 1000)];
end
hist(profitvec, -200:25:600); axis tight
```

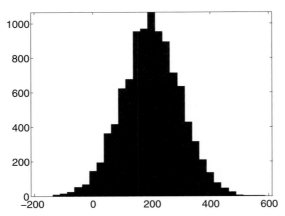

It seems that turning a profit after 10,000 games is highly likely. But though the chance of a loss is quite small at this point, it is not negligible; more than 1% of the trials resulted in a loss, and sometimes the loss was more than 100 units. However, the overall trend toward profitability seems clear, and we expect that after 100,000 games the casino is overwhelmingly likely to have made a profit. On the basis of our previous observations of the growth of the spread of outcomes, we expect that most of the time the net profit will be within 1000 of the expected value of 2000. To keep the computation brief, we'll do only 1000 trials this time:

```
profitvec = [];
for j = 1:10
    profitvec = [profitvec profits(100000, 100)];
end
hist(profitvec, 500:100:3500); axis tight
```

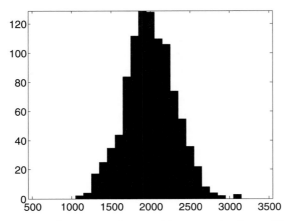

The results are consistent with our projections. The casino seems almost certain to have made a profit after 100,000 games, but it should have reserves of several hundred betting units on hand in order to cover the possible losses along the way.

IQ Sampling: Are You the Smartest Person in the Room?

Now for our second example, suppose that you have an IQ of 130, and that IQs are distributed randomly according to a normal or Gaussian distribution, with mean value 100 and 15 as the standard deviation. If you find yourself in a room full of *random* people, what is the chance that you are the *smartest* person in the room, in the sense of having the highest IQ? We'll use a few commands from the Statistics Toolbox to do some simulations.

Using the Command `random` from the Statistics Toolbox

Let's test this with simulations of a circle of five friends, a classroom of 30 students and the 435-member House of Representatives. Of course, the simulations will only be meaningful if the groups are indeed random. IQ distributions are very likely to be random among friends and classmates; as for Congressmen, there are those who would swear that the distribution is skewed lower than normal – but we'll ignore that objection.

The M-file **intell.m** runs n trials with a cohort of k people. It uses the command **random** from the Statistics Toolbox to generate an $n \times k$ matrix of independent normally distributed IQ values, and then counts to see how many rows (trials) have every entry less than 130. In each case, we'll run 10 trials, then 100 trials, and finally 1000 trials. Here's the code for each trial. (Normal distributions can also be simulated using **randn**, which you can investigate via online help.)

```
type intell
```

```
function frac = intell(n, k)
% computes likelihood of a 130 IQ person being the smartest in
% the room containing k people, repeats this n times, and
% outputs the result.
A = random('norm', 100, 15, [n k]); % n x k random matrix
% The first entry to *|random|* tells Matlab to use the normal
% distribution; Type help random to see other choices.
B = sum(A > 130, 2); % counts no. of entries in each row > 130
frac = sum(B == 0)/n; % counts what fraction of row is zero
```

Circle of Friends

In a circle with five random friends, it's not so hard to be the smartest. Here are our simulations:

```
[intell(10, 5), intell(100, 5), intell(1000, 5)]
```

```
ans =

    1.0000    0.8900    0.8900
```

Repeated runs of 1000 trials show that your likelihood of being the smartest among your friends is about 89%.

A Classroom

It's somewhat harder in a classroom of 30 random students. Of course, it would be harder still in a Lake Wobegon classroom where "all the children are above average" (with apologies to Garrison Keillor). This time you need to beat 29 other students:

```
[intell(10, 29), intell(100, 29), intell(1000, 29)]

ans =

    0.7000      0.5100      0.5070
```

In the classroom, perhaps surprisingly, the probability drops to about 51%.

Congress

It would be still harder in Congress with 435 representatives, where you need to beat 434 other people. Of course, remember we are assuming this is a random sample of the overall population, which might not be a good assumption – and we leave it to you to decide which way the data might be skewed:

```
[intell(10, 434), intell(100, 434), intell(1000, 434)]

ans =

    0       0       0
```

Alas, it is virtually certain that you are not the smartest person in the chamber.

What Does Probability Theory Say?

Given the assumptions above, the fraction of the population with IQs less than yours is the cumulative distribution function of the normal distribution with $\mu = 100$ and $\sigma = 15$, evaluated at 130. This comes out to:

```
p = normcdf(130, 100, 15)

p =

    0.9772
```

Thus, even though you are smarter than roughly 97.7% of the folks, you cannot be assured of being the smartest person in a room unless the size of the crowd is fairly limited. More precisely, in a room with five, 29 and 434 other people, respectively, your chances of being the smartest are

```
[p^5, p^29, p^434]
```

```
ans =

    0.8913    0.5131    0.0000
```

that is, roughly 89%, 51%, and essentially nil. This agrees precisely with the results we saw in the Monte Carlo simulation above.

Population Dynamics

We are going to look at two models for population growth of a species. The first is a standard exponential growth/decay model that describes quite well the population of a species becoming extinct, or the short-term behavior of a population growing in an unchecked fashion. The second, more realistic, model describes the growth of a species subject to constraints of space, food supply, and competitors/predators.

Exponential Growth/Decay

We assume that the species starts with an initial population P_0. The population after n time units is denoted P_n. Suppose that, in each time interval, the population increases or decreases by a fixed proportion of its value at the beginning of the interval. Thus,

$$P_{n+1} = P_n + rP_n, \qquad n \geq 0.$$

The constant r represents the difference between the birth rate and the death rate. The population increases if r is positive, decreases if r is negative, and remains fixed if $r = 0$.

Here is a simple M-file that we will use to compute the population at stage n, given the population at the previous stage and the rate r:

type itseq

```
function X = itseq(f, Xinit, n, r)
% Computes an iterative sequence of values.
X = zeros(n + 1, 1);
X(1) = Xinit;
for k = 1:n
    X(k + 1) = f(X(k), r);
end
```

In fact, this is a simple program for computing iteratively the values of a sequence $x_{k+1} = f(x_k, r), n \geq 0$, given any function f, the value of its parameter r, and the initial value x_0 of the sequence.

Now let's use the program to compute two populations at five-year intervals for $r = 0.1$ and then $r = -0.1$:

```
Xinit = 100; f = @(x, r) x*(1 + r);
X = itseq(f, Xinit, 100, 0.1);
format long; X(1:5:101)

ans =

   1.0e+06 *

   0.000100000000000
   0.000161051000000
   0.000259374246010
   0.000417724816942
   0.000672749994933
   0.001083470594339
   0.001744940226889
   0.002810243684806
   0.004525925556818
   0.007289048368510
   0.011739085287970
   0.018905914247128
   0.030448163954142
   0.049037072529785
   0.078974695679944
   0.127189537139507
   0.204840021458548
   0.329896902959206
   0.531302261184831
   0.855667604660783
   1.378061233982238


X = itseq(f, Xinit, 100, -0.1); X(1:5:101)

ans =

   1.0e+02 *

   1.000000000000000
   0.590490000000000
   0.348678440100000
   0.205891132094649
   0.121576654590569
   0.071789798769185
   0.042391158275216
   0.025031555049932
   0.014780882941435
   0.008727963568088
   0.005153775207320
   0.003043252722170
   0.001797010299914
   0.001061116611996
```

```
0.000626578748218
0.000369988485035
0.000218474500528
0.000129007007817
0.000076177348046
0.000044981962248
0.000026561398888
```

In the first case, the population is growing rapidly; in the second, decaying rapidly. In fact, it is clear from the model that, for any n, the quotient $P_n/P_{n+1} = (1+r)$, and therefore it follows that $P_n = P_0(1+r)^n, n \geq 0$. This accounts for the expression "exponential growth/decay." The model predicts a population growth without bound (for growing populations), and is therefore not realistic. Our next model allows for a check on the population caused by limited space, limited food supply, competitors, and predators.

Logistic Growth

The previous model assumes that the relative change in population is constant, that is

$$(P_{n+1} - P_n)/P_n = r.$$

Now let's build in a term that holds down the growth, namely

$$(P_{n+1} - P_n)/P_n = r - uP_n.$$

We shall simplify matters by assuming that $u = 1 + r$, so that our recursion relation becomes

$$P_{n+1} = uP_n(1 - P_n),$$

where u is a positive constant. In this model, the population P is constrained to lie between 0 and 1, and should be interpreted as a percentage of a maximum possible population in the environment in question. So let us define the function we will use in the iterative procedure:

```
f = @(x, u) u*x*(1 - x);
```

Now let's compute a few examples, and use **plot** to display the results:

```
Xinit = 0.5; X = itseq(f, Xinit, 20, 0.5); plot(X)
```

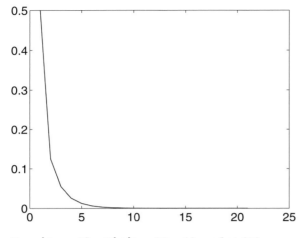

`X = itseq(f, Xinit, 20, 1); plot(X)`

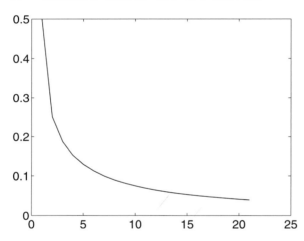

`X = itseq(f, Xinit, 20, 1.5); plot(X)`

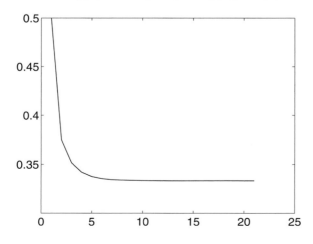

```
X = itseq(f, Xinit, 20, 3.4); plot(X)
```

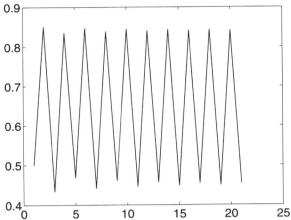

In the first computation, we have used our iterative program to compute the population density for 20 time intervals, assuming a logistic growth constant $u = 0.5$, and an initial population density of 50%. The population seems to be dying out. In the remaining examples, we kept the initial population density at 50%; the only thing we varied was the logistic growth constant. In the second example, with a growth constant $u = 1$, once again the population is dying out – although more slowly. In the third example, with a growth constant of 1.5, the population seems to be stabilizing at 33.3...%. Finally, in the last example, with a constant of 3.4, the population seems to oscillate between densities of approximately 45% and 84%.

These examples illustrate the remarkable features of the logistic population dynamics model. This model has been studied for more than 150 years, its origins lying in an analysis by the Belgian mathematician Pierre-François Verhulst. Here are some of the facts associated with this model. We will corroborate some of them with MATLAB. In particular, we shall use **bar** as well as **plot** to display some of the data.

- **The logistic constant cannot be larger than** 4.

In order for the model to work, the output at any point must be between 0 and 1. But the parabola $ux(1-x)$, for $0 \leq x \leq 1$, has its maximum height when $x = 1/2$, where its value is $u/4$. To keep that number between 0 and 1, we must restrict $u \leq 4$. Here is what happens if u is greater than 4:

```
X = itseq(f, 0.5, 10, 4.5)

X =

   1.0e+173 *

    0.000000000000000
    0.000000000000000
   -0.000000000000000
   -0.000000000000000
```

```
-0.000000000000000
-0.000000000000000
-0.000000000000000
-0.000000000000000
-0.000000000000000
-0.000000000000000
-8.315065427135959
```

- **If $0 \le u \le 1$, the population density tends to zero for any initial value.**

```
X = itseq(f, 0.5, 100, 0.8); X(101)
```

```
ans =
```

```
    2.420473970178059e-11
```

```
X = itseq(f, 0.5, 20, 1); bar(X)
```

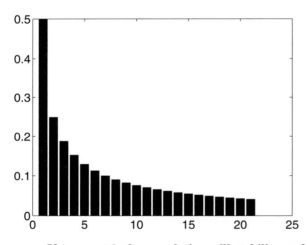

- **If $1 < u \le 3$, the population will stabilize at density $1 - 1/u$ for any initial density other than zero.**

The third of the original four examples corroborates the assertion (with $u = 1.5$ and $1 - 1/u = 1/3$). In the following examples, we set $u = 2, 2.5,$ and 3, respectively, so that $1 - 1/u$ equals $0.5, 0.6,$ and $0.666\ldots$, respectively. The convergence in the last computation is rather slow (as might be expected from a boundary case – or "bifurcation point").

```
X = itseq(f, 0.25, 100, 2); X(101)
```

```
ans =
```

```
    0.500000000000000
```

```
X = itseq(f, 0.75, 100, 2); X(101)
```

```
ans =
```

```
   0.50000000000000
```

```
X = itseq(f, 0.5, 20, 2.5);
plot(X)
```

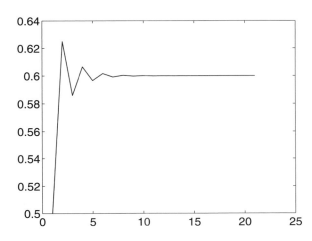

```
X = itseq(f, 0.5, 100, 3);
bar(X); axis([0 100 0 0.8])
```

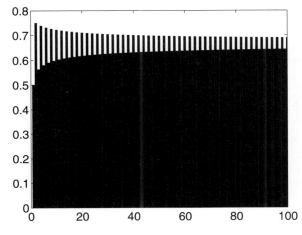

- **If** $3 < u < 3.56994\ldots$, **then there is a periodic cycle.**

The theory is quite subtle. For a fuller explanation, the reader may consult *Encounters with Chaos*, by Denny Gulick, McGraw-Hill, New York, 1992, Section 1.5. In fact, there is a sequence

$$u_0 = 3 < u_1 = 1 + \sqrt{6} < u_2 < u_3 < \cdots < 4,$$

such that between u_0 and u_1 there is a cycle of period 2; between u_1 and u_2 there is cycle of period 4; and, in general, between u_k and u_{k+1} there is a cycle of period 2^{k+1}. In fact, at least for small k, we have the approximation $u_{k+1} \approx 1 + \sqrt{3 + u_k}$. So:

```
u1 = 1 + sqrt(6)

u1 =

    3.449489742783178


u2approx = 1 + sqrt(3 + u1)

u2approx =

    3.539584561061746
```

This explains the oscillatory behavior we saw in the last of the original four examples (with $u_0 < u = 3.4 < u_1$). Here is the behavior for $u_1 < u = 3.5 < u_2$. The command **bar** is particularly effective here for spotting the cycle of order 4.

```
X = itseq(f, 0.75, 100, 3.5);
bar(X); axis([0 100 0 0.9])
```

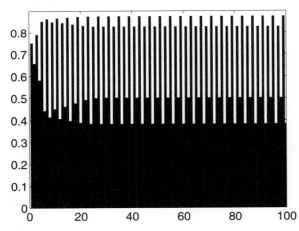

- **There is a value $u < 4$ beyond which – chaos!**

It is possible to prove that the sequence u_k tends to a limit u_∞. The value of u_∞, sometimes called the "Feigenbaum parameter," is approximately $3.56994\ldots$. Let's see what happens if we use a value of u between the Feigenbaum parameter and 4.

```
X = itseq(f, 0.75, 100, 3.7);
plot(X); axis([0 100 0 1])
```

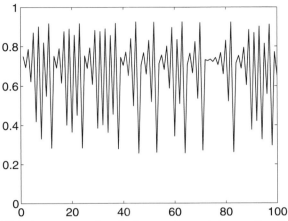

This is an example of what mathematicians call a "chaotic" phenomenon! It is not random – the sequence was generated by a precise, fixed mathematical procedure, but the results manifest no discernible pattern. Chaotic phenomena are unpredictable, but with modern methods (including computer analysis), mathematicians have been able to identify certain patterns of behavior in chaotic phenomena. For example, the last figure suggests the possibility of unstable periodic cycles and other recurring phenomena. Indeed, a great deal of information is known. The aforementioned book by Gulick is a fine reference, as well as a source of an excellent bibliography on the subject.

Re-running the Model with Simulink

The logistic growth model that we have been exploring lends itself particularly well to simulation using Simulink. Here is a simple Simulink model that corresponds to the above calculations:

open_system popdy

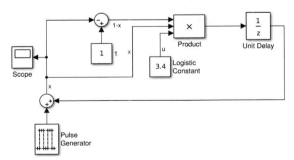

Let's briefly explain how this works. If you ignore the Pulse Generator block and the Sum block in the lower left for a moment, this model implements the equation

$$x \text{ at next time} = ux(1 - x) \text{ at current time,}$$

which is the equation for the logistic model. The Scope block displays a plot of x as a function of (discrete) time. However, we need somehow to build in the initial condition for x. The simplest way to do this is as illustrated here: we add to the right-hand side a discrete pulse that is the initial value of x (here we use 0.5) at time $t = 0$ and is 0 thereafter. Since the model is discrete, you can achieve this by setting the Pulse Generator block to "Sample based" mode, setting the period of the pulse to something longer than the length of the simulation, setting the width of the pulse to 1, and setting the amplitude of the pulse to the initial value of x. The outputs from the model in the two interesting cases of $u = 3.4$ and $u = 3.7$ are shown here:

```
set_param('popdy/Logistic Constant', 'Value', '3.4')
popout = sim('popdy', 'SaveTime','on','TimeSaveName','t', ...
    'SaveState','on','StateSaveName','x','StopTime','120');
t = popout.find('t'); x = popout.find('x');
simplot(t, x); title('u = 3.4')
set(gcf, 'InvertHardcopy', 'off')
```

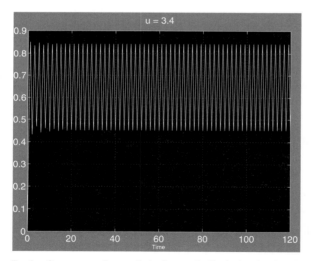

In the first case of $u = 3.4$, the periodic behavior is clearly visible.

```
set_param('popdyn/Logistic Constant', 'Value', '3.7')
popout = sim('popdy', 'SaveTime','on','TimeSaveName','t', ...
    'SaveState','on','StateSaveName','x','StopTime','120');
t = popout.find('t'); x = popout.find('x');
simplot(t, x); title('u = 3.7')
set(gcf, 'InvertHardcopy', 'off')
```

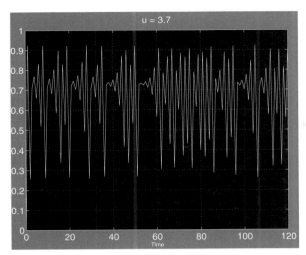

On the other hand, when $u = 3.7$, we get chaotic behavior.

Linear Programming

MATLAB is ideally suited to handle so-called linear programming problems. These are problems in which you have a quantity, depending linearly on several variables, that you want to maximize or minimize subject to several constraints that are expressed as linear inequalities in the same variables. If the number of variables and the number of constraints are small, then there are numerous mathematical techniques for solving a linear programming problem – indeed, these techniques are often taught in high-school or university-level courses in finite mathematics. But sometimes these numbers are high, or, even if they are low, the constants in the linear inequalities or the object expression for the quantity to be optimized may be numerically complicated – in which case a software package like MATLAB is required to effect a solution. We shall illustrate the method of linear programming by means of a simple example, giving a combined graphical/numerical solution, and then solve both a slightly as well as a substantially more complicated problem.

Suppose that a farmer has 75 acres on which to plant two crops: wheat and barley. To produce these crops, it costs the farmer (for seed, fertilizer, etc.) $120 per acre for the wheat and $210 per acre for the barley. The farmer has $15,000 available for expenses. But after the harvest, the farmer must store the crops while awaiting favorable market conditions. The farmer has storage space for 4000 bushels. Each acre yields an average of 110 bushels of wheat or 30 bushels of barley. If the net profit per bushel of wheat (after all expenses have been subtracted) is $1.30 and for barley is $2.00, how should the farmer plant the 75 acres to maximize profit?

We begin by formulating the problem mathematically. First, we express the objective, that is the profit, and the constraints algebraically, then we graph them, and lastly we arrive at the solution by graphical inspection and a minor arithmetic calculation.

Let x denote the number of acres allotted to wheat and y the number of acres allotted to barley. Then the expression to be maximized, that is the profit, is clearly

$$P = (110)(1.30)x + (30)(2.00)y = 143x + 60y.$$

There are three constraint inequalities, specified by the limits on expenses, storage, and acreage. They are, respectively,

$$120x + 210y \le 15000,$$

$$110x + 30y \le 4000,$$

$$x + y \le 75.$$

Strictly speaking there are two more constraint inequalities forced by the fact that the farmer cannot plant a negative number of acres, namely

$$x \ge 0, \quad y \ge 0.$$

Next we graph the regions specified by the constraints. The last two say that we need to consider only the first quadrant in the x-y plane. Here's a graph delineating the triangular region in the first quadrant determined by the first inequality:

```
X = 0:125;
Y1 = (15000 - 120.*X)./210;
area(X, Y1)
```

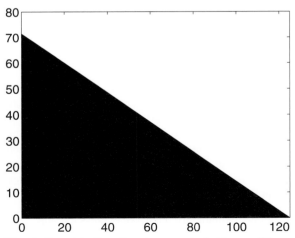

Now let's put in the other two constraint inequalities:

```
Y2 = max((4000 - 110.*X)./30, 0);
Y3 = max(75 - X, 0);
Ytop = min([Y1; Y2; Y3]);
area(X, Ytop)
```

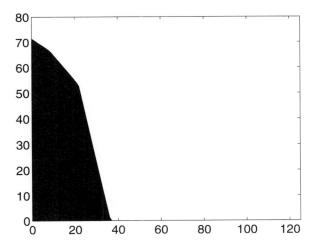

It's a little hard to see the polygonal boundary of the region clearly. Let's home in a bit:

```
area(X, Ytop); axis([0 40 40 75])
```

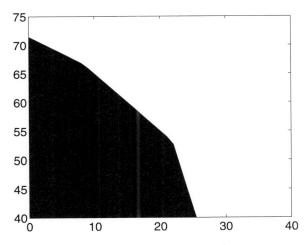

Now let's superimpose on top of this picture a contour plot of the objective function P:

```
hold on
[U V] = meshgrid(0:40, 40:75);
contour(U, V, 143.*U + 60.*V); hold off
```

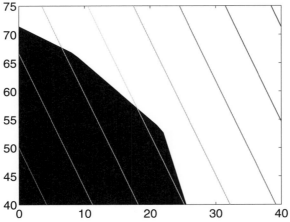

It seems apparent that the maximum value of P will occur on the level curve (that is, level line) that passes through the vertex of the polygon that lies near $(22, 53)$. In fact, we can compute:

```
[x, y] = solve('x + y = 75', '110*x + 30*y = 4000')
```

```
x =

175/8
```

```
y =

425/8
```

```
double([x, y])
```

```
ans =

   21.8750    53.1250
```

The acreage that results in the maximum profit is 21.875 for wheat and 53.125 for barley. In that case, the profit is:

```
P = 143*x + 60*y
```

```
P =

50525/8
```

format bank; double(P)

```
ans =

      6315.62
```

that is, $6,315.63.

This problem illustrates and is governed by the "Fundamental Theorem of Linear Programming," which is stated here for two variables: a linear expression $ax + by$, defined over a closed bounded convex set S whose sides are line segments, takes on its maximum and minimum values at vertices of S. If S is unbounded, there might but need not be an optimum value, but, if there is, it occurs at a vertex. (A convex set is one for which any line segment joining two points of the set lies entirely inside the set.)

In fact, the Optimization Toolbox has a built-in function, **linprog**, which implements the solution of a linear programming problem. You can learn about it from the online help. If you don't have access to the Optimization Toolbox, there is a similar function **simlp** in Simulink. We will use **linprog** on the above problem. After that, we will use it to solve two more complicated problems involving more variables and constraints. Here is the beginning of its help text:

helptext = help('linprog'); helptext(1:190)

```
ans =

  LINPROG Helper function for GETX0; solves linear programming problem.
     X=LINPROG(f,A,b) solves the linear programming problem:

            min f'x     subject to:   Ax <= b
              x
```

So:

f = [-143 -60];
A = [120 210; 110 30; 1 1; -1 0; 0 -1];
b = [15000; 4000; 75; 0; 0];

format short; linprog(f, A, b)

```
ans =

   21.8750
   53.1250
```

This is the same answer that we obtained before. Note that we entered the negative of the coefficients for the objective function P into the vector **f** because **linprog** searches for a minimum rather than a maximum. Note also that the non-negativity constraints are accounted for in the last two rows of **A** and **b**.

Well, we could have done this problem by hand. But suppose that the farmer is dealing with a third crop, say corn, and that the corresponding data are:

- Cost per acre: $150.75,
- Yield per acre: 125 bushels,
- Profit per bushel: $1.56.

If we denote the number of acres allotted to corn by z, then the objective function becomes

$$P = (110)(1.30)x + (30)(2.00)y + (125)(1.56) = 143x + 60y + 195z,$$

and the constraint inequalities are

$$120x + 210y + 150.75z \leq 15000,$$
$$110x + 30y + 125z \leq 4000,$$
$$x + y + z \leq 75,$$
$$x \geq 0, \quad y \geq 0, \quad z \geq 0.$$

The problem is solved with **linprog** as follows:

```
f = [-143 -60 -195];
A = [120 210 150.75; 110 30 125; 1 1 1; -1 0 0; ...
     0 -1 0; 0 0 -1];
b = [15000; 4000; 75; 0; 0; 0];
linprog(f, A, b)

ans =

        0
  56.5789
  18.4211
```

So the farmer should ditch the wheat and plant 56.5789 acres of barley and 18.4211 acres of corn.

There is no practical limit on the number of variables and constraints that MATLAB can handle – certainly none that the relatively unsophisticated user will encounter. Indeed, in many real applications of the technique of linear programming, you may need to deal with many variables and constraints. The solution of such a problem by hand is not feasible, and software like MATLAB is crucial to success. For example, in the farming problem with which we have been working, you could have more crops than two or three – think agribusiness instead of family farmer. Also, you could have constraints that arise from other things beside expenses, storage, and acreage limitations – for example the following.

- Availability of seed. This might lead to constraint inequalities such as $x_j \leq k$.
- Personal preferences. The farmer's spouse might have a preference for one variety over another and insist on a corresponding planting, or something similar with a collection of crops; thus, constraint inequalities such as $x_i \leq x_j$ or $x_1 + x_2 \geq x_3$.
- Government subsidies. It may take a moment's reflection on the reader's part, but this could lead to inequalities such as $x_j \geq k$.

Below is a sequence of commands that solves exactly such a problem. You should be able to recognize the objective expression and the constraints from the data that are entered. But as an aid, you might answer the following questions.

- How many crops are under consideration?
- What are the corresponding expenses? How much is available for expenses?
- What are the yields in each case? What is the storage capacity?
- How many acres are available?
- What crops are constrained by seed limitations? To what extent?
- What about preferences?
- What are the minimum acreages for each crop?

```
f = [-110*1.3 -30*2.0 -125*1.56 -75*1.8 -95*.95 ...
     -100*2.25 -50*1.35];
A = [120 210 150.75 115 186 140 85; ...
     110 30 125 75 95 100 50; 1 1 1 1 1 1 1; 1 0 0 0 0 0 0; ...
     0 0 1 0 0 0 0; 0 0 0 0 0 1 0; 1 -1 0 0 0 0 0; ...
     0 0 1 0 -2 0 0; 0 0 0 -1 0 -1 1; -1 0 0 0 0 0 0; ...
     0 -1 0 0 0 0 0; 0 0 -1 0 0 0 0; 0 0 0 -1 0 0 0; ...
     0 0 0 0 -1 0 0; 0 0 0 0 0 -1 0; 0 0 0 0 0 0 -1];
b = [55000; 40000; 400; 100; 50; 250; 0; 0; 0; -10; -10; ...
     -10; -10; -20; -20; -20];
linprog(f, A, b)

ans =

    10.0000
    10.0000
    40.0000
    45.6522
    20.0000
   250.0000
    20.0000
```

Note that, despite the complexity of the problem, MATLAB solves it almost instantaneously. It seems the farmer should bet the farm on crop number 6. We suggest that you alter the expense and/or the storage limit in the problem and see the effect on the answer.

Image Compression and Manipulation

In Chapter 5, we discussed using MATLAB to read, write, display, and manipulate image data.

In the first part of this section, we illustrate a general technique for data compression based on a tool from linear algebra called "singular value decomposition." A variety of names are associated with this technique, including "principal component analysis," "proper orthogonal decomposition," and "empirical orthogonal functions." In scientific applications, the purpose of the technique is often to identify significant patterns in large data sets and focus computational or visualization techniques on these patterns. The compression example below is meant to illustrate visually how most of the perceptible information in a data set can sometimes be contained in a much smaller data set.

At the end of this section, we give some examples of commands, mostly from the Image Processing Toolbox, to adjust the appearance of an image.

Singular Value Decomposition and Compression

The command **svd** computes the singular value decomposition of a matrix; here we will use the "economy" form of the output (see the online help for other options). While the discussion below is for a general matrix A, we will illustrate with a specific example in MATLAB:

```
A = [1 2 3 4; 5 6 7 8; 9 1 2 3];
[U, S, V] = svd(A, 'econ')

U =

   -0.3135    -0.3175     0.8949
   -0.8067    -0.4081    -0.4274
   -0.5009     0.8559     0.1282


S =

   16.0147          0          0
         0     6.4657          0
         0          0     0.8516


V =

   -0.5530     0.8267    -0.1038
   -0.3727    -0.3446    -0.7588
   -0.4739    -0.3244    -0.0592
   -0.5751    -0.3043     0.6403
```

To explain what the singular value decomposition is, let n be the smaller of the number of rows in A and the number of columns in A. Then the three output matrices have the following properties:

- `A = U*S*V';`

- `S` is an $n \times n$ diagonal matrix; its diagonal entries are, in decreasing order, the "singular values" of A;

- `U'*U = V'*V = eye(n)` (for each matrix, its columns are an orthonormal set of vectors).

As a consequence of the first two properties, we can decompose A as a sum `A = U(:,1)*S(1,1)*V(:,1)' + U(:,2)*S(2,2)*V(:,2)' + ... + U(:,n)*S(1,n)*V(:,n)'`. (Recall that in MATLAB, `U(:,j)` is column j of U.) Let's check this identity for our example:

```
B = zeros(size(A));
for j = 1:3
    B = B + U(:,j)*S(j,j)*V(:,j)';
end
B


B =

    1.0000    2.0000    3.0000    4.0000
    5.0000    6.0000    7.0000    8.0000
    9.0000    1.0000    2.0000    3.0000
```

If A is $n \times k$ where $k \geq n$, then each term in the sum is an $n \times k$ matrix, but it can be computed from only $n + k$ numbers: the n entries in `U(:,j)*S(j,j)` and the k entries in `V(:,j)`.

One can show that the singular values of A are the square roots of the eigenvalues of `A*A'`. Often the singular values of A vary greatly in size, especially if A is large. Since the columns of U and V are unit vectors, the size of `U(:,j)*S(j,j)*V(:,j)'` is determined by `S(j,j)`, and thus the summation above is dominated by the terms with the largest singular values. If we approximate the sum by its r largest terms, we can store the information in the sum using $(n + k)r$ numbers. This is much smaller than the $n * k$ numbers in A, provided r is much smaller than n.

Since each term in the sum is a matrix with rank 1, the sum of the first r terms has rank (at most) r. This sum is in fact the best rank r approximation to A in an appropriate least-squares sense.

Let's look at the rank 1 approximation to our 3×4 example:

```
U(:,1)*S(1,1)*V(:,1)'
```

```
ans =

    2.7762    1.8710    2.3792    2.8874
    7.1438    4.8145    6.1223    7.4301
    4.4362    2.9897    3.8019    4.6140
```

The result doesn't look much like A. Let's try the rank 2 approximation, using a short-cut formula:

```
U(:,1:2)*S(1:2,1:2)*V(:,1:2)'

ans =

    1.0791    2.5783    3.0451    3.5120
    4.9622    5.7238    6.9784    8.2330
    9.0113    1.0828    2.0065    2.9301
```

This is a pretty good approximation, which follows from the fact that the third singular value of our A is so much smaller than the other two.

Compression Example

We now load a sample photograph into MATLAB with **imread** and check its size (we will display the photograph later, after converting the data to shades of gray):

```
photoarray = imread('sample.jpg');
size(photoarray)

ans =

        1536        2048           3
```

The values in the array are red, green, and blue intensities (integer values from 0 to 255) for each of 1536*2048 pixels.

Next, we separate the colors and convert to floating-point values between 0 and 1. We use single precision because it is sufficient for our calculations, and our objective is to store efficiently the essential data in the image. (We are still converting each byte of the integer image data into four bytes of floating-point data, but the floating-point is necessary for the calculations we do below.)

```
redpix = single(photoarray(:,:,1))/255;
greenpix = single(photoarray(:,:,2))/255;
bluepix = single(photoarray(:,:,3))/255;
```

To illustrate the ideas below more simply, we convert to a black-and-white (more precisely, "grayscale") image. The following is a standard formula to convert RGB intensities to gray intensities:

```
graypix = 0.299*redpix + 0.587*greenpix + 0.114*bluepix;
```

Since we have converted to floating-point values, we use **imagesc** rather than **image** to display the photograph. It shows part of Owachomo Bridge in Utah, and was taken by one of the authors. We use **colormap** to get a grayscale display, and **axis** to turn off axis labels and ensure the correct aspect ratio.

```
imagesc(graypix)
colormap(gray), axis off image
```

Now we compute the singular value decomposition of the grayscale data:

```
[U,S,V] = svd(graypix);
```

To see how fast they decay, we graph the first 100 singular values. We open a new figure window so that the photograph remains visible if this M-file is used interactively:

```
figure
plot(diag(S(1:100,1:100)))
```

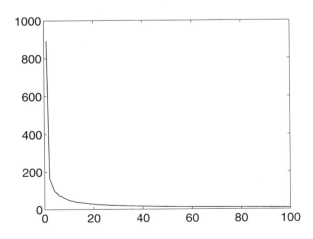

The following function computes and displays the rank n approximation to our image:

```
showrank = @(n) imagesc(U(:,1:n)*S(1:n,1:n)*V(:,1:n)');
```

Below we show the approximations of rank 1, 3, 10, and 30:

```
colormap(gray)
rankvals = [1 3 10 30];
set(0,'DefaultAxesFontSize', 10)
for m=1:4
    subplot(2,2,m)
    showrank(rankvals(m)), axis off image
    title(['Rank ' num2str(rankvals(m))])
end
set(0,'DefaultAxesFontSize', 16)
```

Rank 1

Rank 3

Rank 10

Rank 30

A rank 1 matrix can only produce a "plaid" image, but by the time we get to rank 10, the image starts to look like a photograph, and by rank 30 it is clear we are looking at a landscape. Next, let's compare the rank 100 approximation to the full-resolution image. We clear the existing figure so that the new image is not shown in a subplot:

```
clf
showrank(100); axis off image
```

The rank 100 approximation is not as sharp as the original, but the loss of quality might not be noticed if the original were not available. As a rough comparison to JPEG compression, let's store the grayscale image as a JPEG file with default parameters and see how big it is:

```
imwrite(graypix, 'graysamp.jpg', 'jpg')
d = dir('graysamp.jpg'); d.bytes
```

```
ans =

     474529
```

The raw image data consist of $1536 \cdot 2048$ intensities that could be stored as 1 byte apiece, for a total of 3 megabytes. Since our rank r approximation requires $(1536 + 2048)r = 3584r$ floating-point numbers, and each of our numbers is stored in single precision (4 bytes), we can compute as follows the rank for which our approximation would require about the same amount of data as the JPEG file:

```
r = d.bytes/3584/4
```

```
r =

   33.1005
```

Of course, the JPEG file uses a much more sophisticated algorithm and has much better quality, but we are at least getting a recognizable image with comparable compression, and with some programming effort we could reduce the number of bytes used per number.

Image Manipulation and the Image Processing Toolbox

In Chapter 5, we showed how to create a mirror image and to crop an image by performing these operations on the array of image data. You can also perform various mathematical operations on the data to enhance or suppress certain features of the image. For example, for our grayscale image `graypix`, with values between 0 and 1, the square-root function brightens the image by making all the intensities closer to 1:

imagesc(sqrt(graypix)), axis off image

The Image Processing Toolbox includes a variety of algorithms that manipulate image data. If you have this toolbox installed, it will appear on the menu when you open a new Help Browser tab (e.g., by typing **doc**); click to get an overview of available commands. Some commands provide shortcuts for operations you could perform with basic MATLAB commands; for example, the command **rgb2gray** will produce almost the same result as the arithmetic commands we used above to convert a color image to grayscale. Other commands perform more sophisticated algorithms, and we give a few examples here.

One of the contrast adjustment commands, **adapthisteq**, will both brighten dark areas of the image and darken light areas; it increases contrast in each part of the image.

imagesc(adapthisteq(graypix)), axis off image

Like the brightened image before, visibility is improved in the shadowy part of the photograph, but unlike the brightened image, the clouds have become darker, and the rock striations have become more pronounced.

The toolbox also has a variety of filters that can "soften" an image; the simplest of these, **imfilter**, will convolve the image with a matrix that you specify. Convolving tends to blur edges in the image, which is sometimes the desired effect. On the other hand, the command **wiener2** (named for Norbert Wiener) reduces graininess in surfaces while trying to maintain sharp edges.

```
imagesc(wiener2(graypix, [9 9])), axis off image
```

Here we have chosen to filter over a 9×9 grid rather than the default 3×3; as always, see the online help for other options to this and the other commands we have used. The result de-emphasizes striations in the rocks, but also blurs the trees and smaller clouds.

Other toolbox commands include algorithms to deblur images; starting with MATLAB 2013a, the newest of these is **imsharpen**.

Processing and Analysis of Financial Data

We will use MATLAB to plot, process, and analyze some data from the financial markets. There are huge quantities of such data, but to make the analysis manageable, we restrict attention here to two sets of numbers from 2012: the closing value each day of the Dow Jones Industrial Average (DJIA), a standard measure of the US stock market, and the daily value of the euro in US dollars, a measure of relative confidence in the US and European economies. The data were downloaded from, and is copyright by, Samuel H. Williamson, "Daily Closing Value of the Dow Jones Average, 1885 to Present," MeasuringWorth, 2012, and the European Central Bank Statistical Data Warehouse, 2012.

Both data sets cover only certain days of the year, namely the days when the markets are open. To compensate for this, we assumed that each market indicator remained constant between the close of business on one market day and the next market day, even if several days (such as a holiday weekend) intervened. Thus, since 2012 was a leap year, each data set consisted of a vector of length 366 (for the 366 days of the year 2012). The data were downloaded and converted to vector form using the **Import Data** button on the MATLAB Desktop, then stored into a `mat` file for future use. We begin by loading the data:

```
load('financial.mat')
```

Our first step is to plot the data. Two lesser-known MATLAB functions are useful here. The function **detrend** subtracts from a data set the best-fitting line, which we plot separately with a dotted line. This makes it easier to see the overall trend and the fluctuations from it. The function **datetick** labels the desired axis with dates or times instead of numbers:

```
plot(1:366, dj, 'k'), title 'Dow Jones average 2012'
hold on
plot(1:366, dj - detrend(dj), ':k')
axis([1,366,12000,14000])
datetick('x','mmm')
hold off
```

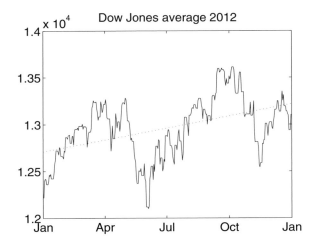

```
figure
plot(1:366, euro, 'k'), title 'Euro/dollar exchange rate 2012'
hold on
plot(1:366, euro - detrend(euro), ':k')
hold off
axis([1,366,1.20,1.35])
datetick('x','mmm')
```

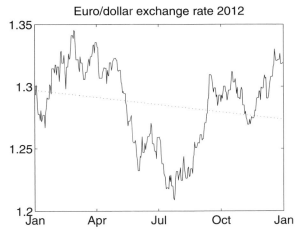

Euro/dollar exchange rate 2012

Now we can begin to analyze the data. For example, we could ask for the correlation between the Dow Jones average and the euro/dollar exchange rate. We might guess that when the US stock market is strong, then the dollar should strengthen against the euro and thus this exchange rate should go down. To test this, we can use **corrcoef**, which computes the correlation coefficient and also gives the p-value, the probability of getting a correlation as large as the observed value by random chance, when the true correlation is zero:

```
[r, p] = corrcoef(dj, euro);
disp(['correlation coefficient = ',num2str(r(1,2))])
disp(['p-value = ',num2str(p(1,2))])
```

```
correlation coefficient = 0.31638
p-value = 5.9494e-10
```

The facts that the correlation is positive and that the p-value is incredibly small indicate that the theory above is certainly false in this case, as the euro strengthened against the dollar even as the Dow Jones average rose.

Next, we observe that if there were no "noise" in the data, we would expect *exponential*, not *linear*, growth or decay in prices. For that reason, it's really better to work with the *logarithms* of the stock exchange index and the euro exchange rate. Let's plot these logarithms with the linear part of the trend subtracted out:

```
ldj = log(dj); ldjdetrend = detrend(ldj);
plot(1:366, ldj, 'k'), title 'log of Dow Jones average 2012'
hold on
plot(1:366, ldj - ldjdetrend, ':k')
axis([1,366,9.2,9.8])
datetick('x','mmm')
hold off
```

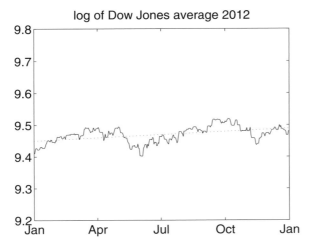

log of Dow Jones average 2012

```
figure
leu = log(euro); leudetrend = detrend(leu);
plot(1:366, leu, 'k')
title 'log of Euro/dollar exchange rate 2012'
hold on
plot(1:366, leu - leudetrend, ':k')
hold off
axis([1,366,0.15,0.30])
datetick('x','mmm')
```

log of Euro/dollar exchange rate 2012

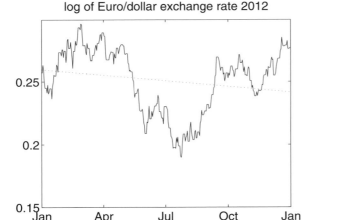

We might also want to know if the data have certain natural frequencies or periodicities. For this, the routine **fft**, which computes the "fast Fourier transform," is useful. Here we want to work with the "detrended" data:

```
y1 = fft(ldjdetrend); y2 = fft(leudetrend);
```

These Fourier transforms are complex numbers. But we can try to plot the real and imaginary parts separately, for example:

```
plot(1:366, real(y1), 'k')
hold on
plot(1:366, imag(y1), ':k')
hold off
```

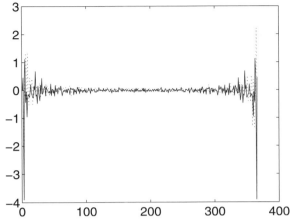

Note that both the solid and dotted plots are very close to zero except for a peak at the beginning and the end. That suggests that the characteristic frequencies of the stock market are all either very small or very large. Let's try the same thing with the exchange rate data:

```
plot(1:366, real(y2), 'k')
hold on
plot(1:366, imag(y2),':k')
hold off
```

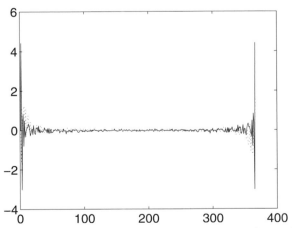

Again, both plots are very close to zero except for peaks at each end.

We may now try to understand the data from the point of view of probability and statistics. For the rest of this application, we use the Statistics Toolbox. First, we can see if both data sets plausibly come from the same distributions, using a quantile-quantile plot constructed with **qqplot**:

```
qqplot(ldjdetrend, leudetrend)
xlabel('Quantiles for detrended log of DJ average')
ylabel('Quantiles for detrended log of euro exchange rate')
```

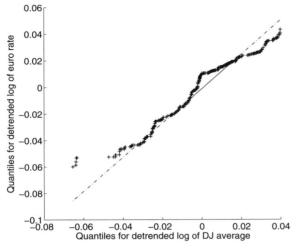

Since this largely follows a straight line (though the tail at one end is a puzzle), we might expect the fluctuations from the linear trend in both data sets to come from the same distribution. The simplest guess would be that the fluctuations roughly follow a normal distribution. (This is the basis for the famous *Black–Scholes model*, which assumes that logs of stock prices, after subtracting off a trend linear in time, are normally distributed.) Let's see if this is plausible, starting say with the DJ data, using a normal probability plot constructed with **normplot**:

normplot(ldjdetrend)

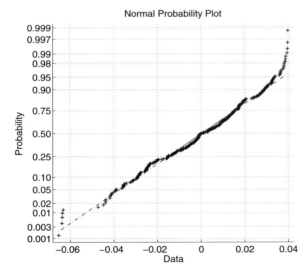

The plot shows that the stock market fluctuations are largely normal, but with "heavy tails" at the ends. In other words, big fluctuations from the mean are more common than in a normal distribution. We should take this into account in applying simple statistical models to the stock market. We can do the same with currency exchange rates:

normplot(leudetrend)

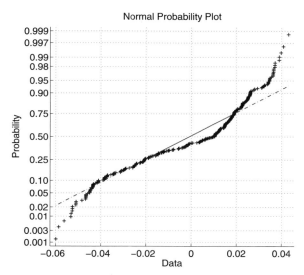

Again we largely have a normal distribution, but with much heavier tails, and not as good a fit as with the Dow Jones average. Now we can try to fit the data to a normal curve with **normfit**:

```
[djm, djsig] = normfit(ldjdetrend);
disp('Parameters for normal fit to the detrended data');
disp(['log of Dow Jones average : mean :', num2str(djm), ...
    ', sigma: ', num2str(djsig)]);
[eurom, eurosig] = normfit(leudetrend);
disp(['log of euro exchange rate : mean :', ...
    num2str(eurom), ', sigma: ', num2str(eurosig)]);
```

```
Parameters for normal fit to the detrended data
log of Dow Jones average : mean :-4.669e-15, sigma: 0.022116
log of euro exchange rate : mean :2.3281e-17, sigma: 0.025166
```

That the means here are negligible is no surprise, as we are working with the detrended data where the trend has already been subtracted. But the standard deviations are large, indicating that big fluctuations in the markets are quite common. More sophisticated analysis of financial data using MATLAB can be done with the Financial Toolbox.

The 360° **Pendulum**

Normally, we think of a pendulum as a weight suspended by a flexible string or cable, so that it may swing back and forth. Another type of pendulum consists of a weight attached by a light (but inflexible) rod to an axle, so that it can swing through larger angles, even making a 360° rotation if given enough velocity.

Equations of Motion

Though it is not precisely correct in practice, we often assume that the magnitude of the frictional forces that eventually slow the pendulum to a halt is proportional to the velocity of the pendulum. Assume also that the length of the pendulum is 1 m, the weight at the end of the pendulum has mass 1 kg, and the coefficient of friction is 0.5. In that case, the equations of motion for the pendulum are

$$x'(t) = y(t), \quad y'(t) = -0.5y(t) - 9.81\sin(x(t)),$$

where t represents time in seconds, x represents the angle of the pendulum from the vertical in radians (so that $x = 0$ is the rest position), y represents the angular velocity of the pendulum in radians per second, and 9.81 is approximately the acceleration due to gravity in meters per second squared. Now here is a phase portrait of the solution with initial position $x(0) = 0$ and initial velocity $y(0) = 5$. This is a graph of x versus y as a function of t, on the time interval $0 < t < 20$. (To use MATLAB's numerical differential equation solver **ode45**, we combine x and y into a single vector **x**; see the online help for **ode45**.)

```
g = @(t, x) [x(2); -0.5*x(2) - 9.81*sin(x(1))];
[t, xa] = ode45(g, [0:0.01:20], [0 5]);
plot(xa(:, 1), xa(:, 2))
```

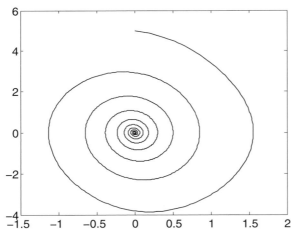

Recall that the x-coordinate corresponds to the angle of the pendulum and the y-coordinate corresponds to its velocity. Starting at $(0, 5)$, as t increases we follow the curve as it spirals clockwise toward $(0, 0)$. The angle oscillates back and forth, but with each swing it gets smaller until the pendulum is virtually at rest by the time $t = 20$. Meanwhile, the velocity oscillates as well, taking its maximum value during each oscillation when the pendulum is in the middle of its swing (the angle is near zero) and crossing zero when the pendulum is at the end of its swing.

A Higher Initial Velocity

We increase the initial velocity to 10:

```
[t, xa] = ode45(g, [0:0.01:20], [0 10]);
plot(xa(:,1), xa(:,2))
```

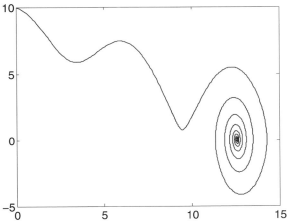

This time the angle increases to over 14 radians before the curve spirals in to a point near $(12.5, 0)$. More precisely, it spirals toward $(4\pi, 0)$, because 4π radians represents the same position for the pendulum as 0 radians does. The pendulum has swung overhead and made two complete revolutions before beginning its damped oscillation toward its rest position. The velocity at first decreases but then rises after the angle passes through π, as the pendulum passes the upright position and gains momentum. The pendulum has just enough momentum to swing through the upright position once more at the angle 3π.

When Does the Pendulum Swing Overhead?

Suppose that we want to find, to within 0.1, the minimum initial velocity required to make the pendulum, starting from its rest position, swing overhead once. It will be useful to be able to see the solutions corresponding to several different initial velocities on one graph.

First we consider the integer velocities 5 to 10:

```
hold on
for a = 5:10
    [t, xa] = ode45(g, [0:0.01:20], [0 a]);
    plot(xa(:, 1), xa(:, 2))
end
hold off
```

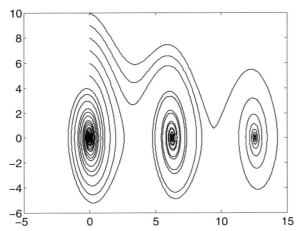

Initial velocities 5, 6, and 7 are not large enough for the angle to increase past π, but initial velocities 8, 9, and 10 are enough to make the pendulum swing overhead. Let's see what happens between 7 and 8:

```
hold on
for a = 7.0:0.2:8.0
    [t, xa] = ode45(g, [0:0.01:20], [0 a]);
    plot(xa(:, 1), xa(:, 2))
end
hold off
```

We see that the cut-off is somewhere between 7.2 and 7.4. Let's make one more refinement:

```
hold on
for a = 7.2:0.05:7.4
    [t, xa] = ode45(g, [0:0.01:20], [0 a]);
    plot(xa(:, 1), xa(:, 2))
end
hold off
```

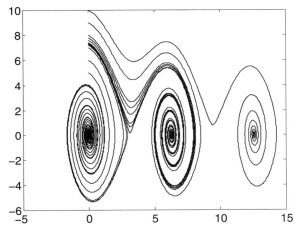

We conclude that the minimum velocity needed is somewhere between 7.25 and 7.3.

☆ Numerical Solution of the Heat Equation

In this section, we will use MATLAB to numerically solve the heat equation (also known as the diffusion equation), a partial differential equation that describes many physical processes such as conductive heat flow or the diffusion of an impurity in a motionless fluid. You can picture the process of diffusion as a drop of dye spreading in a glass of water. (To a certain extent, you could also picture cream in a cup of coffee, but in that case the mixing is generally complicated by the fluid motion caused by pouring the cream into the coffee, and is further accelerated by stirring the coffee.) The dye consists of a large number of individual particles, each of which repeatedly bounces off the surrounding water molecules, following an essentially random path. There are so many dye particles that their individual random motions form an essentially deterministic overall pattern as the dye spreads evenly in all directions (we ignore here the possible effect of gravity). In a similar way, you can imagine heat energy spreading through random interactions of nearby particles.

In a three-dimensional medium, the heat equation is

$$\frac{\partial u}{\partial t} = k \left(\frac{\partial^2 u}{\partial x^2} + \frac{\partial^2 u}{\partial y^2} + \frac{\partial^2 u}{\partial z^2} \right).$$

Here u is a function of t, x, y, and z that represents the temperature, or concentration of impurity in the case of diffusion, at time t at position (x, y, z) in the medium. The constant k depends on the materials involved, and is called the thermal conductivity in the case of heat flow, and the diffusion coefficient in the case of diffusion. To simplify matters, let us assume that the medium is instead one-dimensional. This could represent diffusion in a thin water-filled tube or heat flow in a thin insulated wire; let us think primarily of the case of heat flow. Then the partial differential equation becomes

$$\frac{\partial u}{\partial t} = k\,\frac{\partial^2 u}{\partial x^2},$$

where $u(x,t)$ is the temperature at time t a distance x along the wire.

A Finite-Difference Solution

To solve this partial differential equation, we need both initial conditions of the form $u(x,0) = f(x)$, where $f(x)$ gives the temperature distribution in the wire at time 0, and boundary conditions at the endpoints of the wire; call them $x = a$ and $x = b$. We choose so-called Dirichlet boundary conditions $u(a,t) = T_a$ and $u(b,t) = T_b$, which correspond to the temperature being held steady at values T_a and T_b at the two endpoints. Though an exact solution is available in this scenario, let us instead illustrate the numerical method of finite differences.

To begin with, on the computer we can keep track of the temperature u only at a discrete set of times and a discrete set of positions x. Let the times be $0, \Delta t, 2\Delta t, \ldots, N\Delta t$, and let the positions be $a, a + \Delta x, \ldots, a + J\Delta x = b$, and let $u_j^n = u(a + j\Delta t, n\Delta t)$. After rewriting the partial differential equation in terms of finite-difference approximations to the derivatives, we get

$$\frac{u_j^{n+1} - u_j^n}{\Delta t} = k\,\frac{u_{j+1}^n - 2u_j^n + u_{j-1}^n}{\Delta x^2}.$$

(These are the simplest approximations we can use for the derivatives, and this method can be refined by using more accurate approximations, especially for the t derivative.) Thus if, for a particular n, we know the values of u_j^n for all j, we can solve the equation above to find for each j,

$$u_j^{n+1} = u_j^n + \frac{k\Delta t}{\Delta x^2}(u_{j+1}^n - 2u_j^n + u_{j-1}^n) = s(u_{j+1}^n + u_{j-1}^n) + (1 - 2s)u_j^n,$$

where $s = k\Delta t/(\Delta x)^2$. In other words, this equation tells us how to find the temperature distribution at time step $n + 1$ given the temperature distribution at time step n. (At the endpoints $j = 0$ and $j = J$, this equation refers to temperatures outside the prescribed range for x, but at these points we will ignore the equation above and apply the boundary conditions instead.) We can interpret this equation as saying that the temperature at a given location at the next time step is a weighted average of its temperature and the temperatures of its neighbors at the current time step. In other words, in time Δt, a given section of the wire of length Δx transfers to each of its neighbors a portion s of its heat energy and keeps the remaining portion $1 - 2s$.

Thus, our numerical implementation of the heat equation is a discretized version of the microscopic description of diffusion we gave initially – that heat energy spreads due to random interactions between nearby particles.

The following M-file, which we have named `heat.m`, iterates the procedure described above.

type heat

```
function u = heat(k, t, x, init, bdry)
% Solve the 1D heat equation on the rectangle described by
% vectors x and t with initial condition u(t(1), x) = init
% and Dirichlet boundary conditions u(t, x(1)) = bdry(1),
% u(t, x(end)) = bdry(2).

J = length(x);
N = length(t);
dx = mean(diff(x));
dt = mean(diff(t));
s = k*dt/dx^2;

u = zeros(N,J);
u(1,:) = init;

for n = 1:N-1
    u(n+1,2:J-1) = s*(u(n,3:J) + u(n,1:J-2)) + ...
        (1-2*s)*u(n,2:J-1);
    u(n+1,1) = bdry(1);
    u(n+1,J) = bdry(2);
end
```

The function **heat** takes as inputs the value of k, vectors of t- and x-values, a vector **init** of initial values (which is assumed to have the same length as **x**), and a vector **bdry** containing a pair of boundary values. Its output is a matrix of u-values. Notice that since indices of arrays in MATLAB must start at 1, not 0, we have deviated slightly from our earlier notation by letting $n = 1$ represent the initial time and $j = 1$ represent the left endpoint. Notice also that, in the first line following the **for** statement, we compute an entire row of **u**, except for the first and last values, in one line; each term is a vector of length $J - 2$, with the index j increased by 1 in the term **u(n,3:J)** and decreased by 1 in the term **u(n,1:J-2)**.

Let's use the M-file above to solve the one-dimensional heat equation with $k = 2$ on the interval $-5 \leq x \leq 5$ from time 0 to time 4, using boundary temperatures 15 and 25, and an initial temperature distribution of 15 for $x < 0$ and 25 for $x > 0$. You can imagine that two separate wires of length 5 with different temperatures are joined at time 0 at position $x = 0$, and each of their far ends remains in an environment that holds it at its initial temperature. We must choose values for Δt and Δx; let's try $\Delta t = 0.1$ and $\Delta x = 0.5$, so that there are 41 values of t ranging from 0 to 4 and 21 values of x ranging from -5 to 5.

```
tvals = linspace(0, 4, 41);
xvals = linspace(-5, 5, 21);
init = 20 + 5*sign(xvals);
uvals = heat(2, tvals, xvals, init, [15 25]);
surf(xvals, tvals, uvals)
xlabel x; ylabel t; zlabel u
```

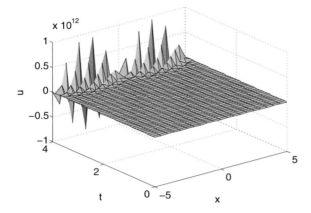

Here we used **surf** to show the entire solution $u(x, t)$. The output is clearly unrealistic; notice the scale on the u-axis! The numerical solution of partial differential equations is fraught with dangers, and instability like that seen above is a common problem with finite-difference schemes. For many partial differential equations, a finite-difference scheme will not work at all, but for the heat equation and similar equations, it will work well with proper choice of Δt and Δx. We might be inclined to think that, since our choice of Δx was larger, it should be reduced, but in fact this would only make matters worse. Ultimately, the only parameter in the iteration we're using is the constant s, and one drawback of doing all the computations in an M-file as we did above is that we do not automatically see the intermediate quantities it computes. In this case, we can easily calculate that $s = 2(0.1)/(0.5)2 = 0.8$. Notice that this implies that the coefficient $1 - 2s$ of u_j^n in the iteration above is negative. Thus, the "weighted average" we described before in our interpretation of the iterative step is not a true average; each section of wire is transferring more energy than it has at each time step!

The solution to the problem above is thus to reduce the time step Δt; for instance, if we cut it in half, then $s = 0.4$, and all coefficients in the iteration are positive:

```
tvals = linspace(0, 4, 81);
uvals = heat(2, tvals, xvals, init, [15 25]);
surf(xvals, tvals, uvals)
xlabel x; ylabel t; zlabel u
```

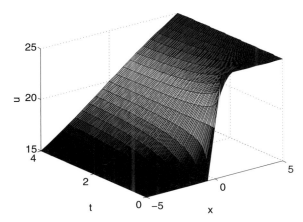

This looks much better! As time increases, the temperature distribution seems to approach a linear function of x. Indeed, $u(x,t) = 20 + x$ is the limiting "steady state" for this problem; it satisfies the boundary conditions and it yields 0 on both sides of the partial differential equation.

Generally speaking, it is best to understand some of the theory of partial differential equations before attempting a numerical solution as we have done here. However, for this particular case at least, the simple rule of thumb of keeping the coefficients of the iteration positive yields realistic results. A theoretical examination of the stability of this finite-difference scheme for the one-dimensional heat equation shows that indeed any value of s between 0 and 0.5 will work, and suggests that the best value of Δt to use for a given Δx is the one that makes $s = 0.25$. Notice that, while we can get more accurate results in this case by reducing Δx, if we reduce it by a factor of 10, we must reduce Δt by a factor of 100 to compensate, making the computation take 1000 times as long and use 1000 times the memory!

The Case of Variable Conductivity

Earlier we mentioned that the problem we solved numerically could also be solved analytically. The value of the numerical method is that it can be applied to similar partial differential equations for which an exact solution is not possible or at least not known. For example, consider the one-dimensional heat equation with a variable coefficient, representing an inhomogeneous material with varying thermal conductivity $k(x)$,

$$\frac{\partial u}{\partial t} = \frac{\partial}{\partial x}\left(k(x)\frac{\partial u}{\partial x}\right) = k(x)\frac{\partial^2 u}{\partial x^2} + k'(x)\frac{\partial u}{\partial x}.$$

For the first derivatives on the right-hand side, we use a symmetric finite-difference approximation, so that our discrete approximation to the partial differential equations becomes

$$\frac{u_j^{n+1} - u_j^n}{\Delta t} = k_j \frac{u_{j+1}^n - 2u_j^n + u_{j-1}^n}{\Delta x^2} + \frac{k_{j+1} - k_{j-1}}{2\Delta x} \frac{u_{j+1} - u_{j-1}}{2\Delta x},$$

where $k_j = k(a + j\Delta x)$. Then the time iteration for this method is

$$u_j^{n+1} = s_j(u_{j+1}^n + u_{j-1}^n) + (1 - 2s_j)u_j^n + 0.25(s_{j+1} - s_{j-1})(u_{j+1}^n - u_{j-1}^n),$$

where $s_j = k_j \Delta t / (\Delta x)^2$. In the following M-file, which we called `heatvc.m`, we modify our previous M-file to incorporate this iteration:

type heatvc

```
function u = heatvc(k, t, x, init, bdry)
% Solve the 1D heat equation with variable-coefficient k
% on the rectangle described by vectors x and t with
% u(t(1), x) = init and Dirichlet boundary conditions
% u(t, x(1)) = bdry(1), u(t, x(end)) = bdry(2).

J = length(x);
N = length(t);
dx = mean(diff(x));
dt = mean(diff(t));
s = k*dt/dx^2;

u = zeros(N,J);
u(1,:) = init;

for n = 1:N-1
    u(n+1,2:J-1) = s(2:J-1).*(u(n,3:J) + u(n,1:J-2)) + ...
        (1 - 2*s(2:J-1)).*u(n,2:J-1) + ...
        0.25*(s(3:J) - s(1:J-2)).*(u(n,3:J) - u(n,1:J-2));
    u(n+1,1) = bdry(1);
    u(n+1,J) = bdry(2);
end
```

Notice that **k** is now assumed to be a vector with the same length as **x**, and that as a result so is **s**. This in turn requires that we use vectorized multiplication in the main iteration, which we have now split into three lines.

Let's use this M-file to solve the one-dimensional variable coefficient heat equation with the same boundary and initial conditions as before, using $k(x) = 1 + (x/5)^2$. Since the maximum value of k is 2, we can use the same values of Δt and Δx as before:

```
kvals = 1 + (xvals/5).^2;
uvals = heatvc(kvals, tvals, xvals, init, [15 25]);
surf(xvals, tvals, uvals)
xlabel x; ylabel t; zlabel u
```

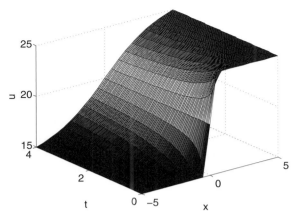

In this case, the limiting temperature distribution is not linear; it has a steeper temperature gradient in the middle, where the thermal conductivity is lower. Again we could find the exact form of this limiting distribution, $u(x, t) = 20(1 + (1/\pi) \arctan(x/5))$, by setting the t derivative to zero in the original equation and solving the resulting ordinary differential equation.

You can use the method of finite differences to solve the heat equation in two or three space dimensions as well. For this and other partial differential equations with time and two space dimensions, you can also use the PDE Toolbox, which implements the more sophisticated "finite-element method."

A Simulink Solution

We can also solve the heat equation using Simulink. To do this, we continue to approximate the x-derivatives with finite differences, but think of the equation as a vector-valued ordinary differential equation, with t as the independent variable. Simulink solves the model using MATLAB's ODE solver, **ode45**. To illustrate how to do this, let's take the example we started with, the case in which $k = 2$ on the interval $-5 \leq x \leq 5$ from time 0 to time 4, using boundary temperatures 15 and 25, and an initial temperature distribution of 15 for $x < 0$ and 25 for $x > 0$. We replace $u(x, t)$ for fixed t by the vector **u** of values of $u(x, t)$, with, say, **x = -5:5**. Here there are 11 values of x at which we are sampling u, but since $u(x, t)$ is predetermined at the endpoints, we can take **u** to be a nine-dimensional vector, and we just tack on the values at the endpoints when we have finished. Since we're replacing $\partial^2/\partial x^2$ by its finite difference approximation and we've taken $\Delta x = 1$ for simplicity, our equation becomes the vector-valued ODE

$$\frac{\partial \mathbf{u}}{\partial t} = k(A\mathbf{u} + c)$$

Here the right-hand side represents our approximation to $k(\partial^2 u/\partial x^2)$. The matrix A is

$$\begin{pmatrix} -2 & 1 & \cdots & 0 \\ 1 & -2 & \ddots & \vdots \\ \vdots & \ddots & \ddots & 1 \\ 0 & \cdots & 1 & -2 \end{pmatrix},$$

since we are replacing $\partial^2 u/\partial x^2$ at (n,t) with $u(n-1,t) - 2u(n,t) + u(n+1,t)$. We represent this matrix in MATLAB's notation by **-2*eye(9) + diag(ones(8,1), 1) + diag(ones(8,1), -1)**. The vector **c** comes from the boundary conditions, and has 15 in its first entry, 25 in its last entry, and 0's in between. We represent it in MATLAB's notation as **[15; zeros(7,1); 25]**. The formula for **c** comes from the fact that **u(1)** represents $u(-4,t)$, and $\partial^2 u/\partial x^2$ at this point is approximated by

$$u(-5,t) - 2u(-4,t) + u(-3,t) = 15 - 2\mathbf{u(1)} + \mathbf{u(2)}$$

and similarly at the other endpoint. Here's a Simulink model representing this equation:

open_system heateq

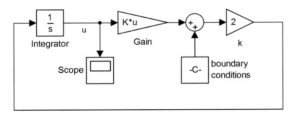

Note that we need to specify the initial conditions for u as Block Parameters for the Integrator block, and that in the Block Parameters dialog box for the Gain block, we need to set the multiplication type to "Matrix". Since **u(1)** through **u(4)** represent the solution $u(x,t)$ at $x = -4$ through -1, and **u(6)** through **u(9)** represent $u(x,t)$ at $x = 1$ through 4, we take the initial value of u to be **[15*ones(4,1); 20; 25*ones(4,1)]**. (We use 20 as a compromise at $x = 0$, since this is right in the middle of the regions where u is 15 and 25.) The output from the model is displayed in the **Scope** block in the form of graphs of the various entries of u as a function of t, but it's more useful to save the output to the MATLAB Workspace and then plot it with **surf**. Incidentally, it helps to reset the y-axis limits on the **Scope** block to run from 15 to 25. To make this adjustment and run the model from $t = 0$ to $t = 4$, we execute the commands:

```
set_param('heateq/Scope', 'YMin', '15');
set_param('heateq/Scope', 'YMax', '25');
hout = sim('heateq', 'SaveTime','on','TimeSaveName','t', ...
    'SaveState','on','StateSaveName','u','StopTime','4');
```

```
tout = hout.find('t'); uout = hout.find('u');
simplot(tout, uout)
set(gcf, 'InvertHardcopy', 'off')
```

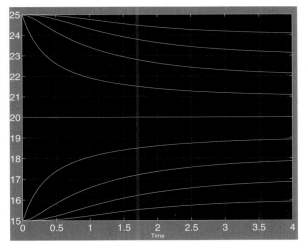

This saves the simulation times (from 0 to 4) as a vector **tout** and the computed values of u as **uout**, a matrix with nine columns. Each row of these arrays corresponds to a single time step, and each column of **uout** corresponds to one value of x. But remember that we have to add in the values of u at the endpoints as additional columns in u. So we plot the data as follows:

```
u = [15*ones(length(tout),1), uout, ...
     25*ones(length(tout),1)];
x = -5:5;
clf reset
set(gcf, 'Color', 'White')
surf(x, tout, u)
xlabel('x'), ylabel('t'), zlabel('u')
title('Solution to heat equation in a rod')
```

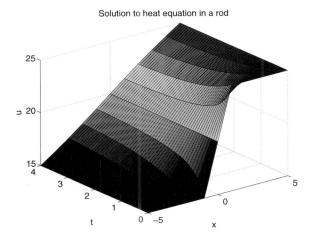

Notice how similar this is to the picture obtained before for constant conductivity $k = 2$. We leave it to the reader to modify the model for the case of variable conductivity.

Solution with pdepe

MATLAB has a built-in solver **pdepe** for partial differential equations in one space dimension (as well as time t). To find out more about it, read the online help on **pdepe**. The instructions for use of **pdepe** are quite explicit but somewhat complicated. The method it uses is somewhat similar to that used in the Simulink solution above; i.e., it uses an ODE solver in t and finite differences in x. The following M-file solves the second problem above, the one with variable conductivity. Note the use of function handles and subfunctions:

type heateqex2

```
function heateqex2
% Solves a Dirichlet problem for the heat equation in a rod,
% this time with variable conductivity, 21 mesh points.
m = 0;  % This simply means geometry is linear.
x = linspace(-5, 5, 21);
t = linspace(0, 4, 81);
sol = pdepe(m, @pdex, @pdexic, @pdexbc, x, t);
% Extract the first solution component as u.
u = sol(:,:,1);
% A surface plot is often a good way to study a solution.
surf(x, t, u)
title('Numerical solution computed with 21 mesh points in x')
xlabel('x'), ylabel('t'), zlabel('u')
% ----------------------------------------------------------
function [c, f, s] = pdex(x, t, u, DuDx)
c = 1;
f = (1 + (x/5).^2)*DuDx;  % flux = conductivity times u_x
s = 0;
% ----------------------------------------------------------
function u0 = pdexic(x)
u0 = 20 + 5*sign(x);  % initial condition at t = 0
% ----------------------------------------------------------
function [pl, ql, pr, qr] = pdexbc(xl, ul, xr, ur, t)
% q's are zero since we have Dirichlet conditions
% pl = 0 at the left, pr = 0 at the right endpoint
pl = ul - 15;
ql = 0;
pr = ur - 25;
qr = 0;
```

Running it gives:

heateqex2

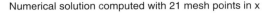

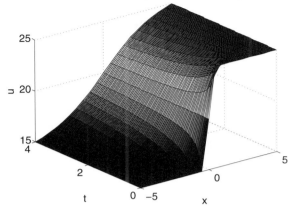

Numerical solution computed with 21 mesh points in x

Again the results are very similar to those obtained before.

☆ A Model of Traffic Flow

Everyone has had the experience of sitting in a traffic jam, or of seeing cars bunch up on a road for no apparent good reason. MATLAB and Simulink are good tools for studying models of such behavior. Our analysis here will be based on so-called "follow-the-leader" theories of traffic flow, about which you can read more in *Kinetic Theory of Vehicular Traffic*, by Ilya Prigogine and Robert Herman, Elsevier, New York, 1971, or in *The Theory of Road Traffic Flow*, by Winifred Ashton, Methuen, London, 1966. We will analyze here an extremely simple model that already exhibits quite complicated behavior. We consider a one-lane, one-way, circular road with a number of cars on it (a very primitive model of, say, the Outer Loop of the Capital Beltway around Washington, DC, since, in very dense traffic, it is hard to change lanes and each lane behaves like a one-lane road). Each driver slows down or speeds up on the basis of his own speed, the speed of the car ahead of him, and the distance to the car ahead of him. But human drivers have a finite reaction time. In other words, it takes them a certain amount of time (usually about a second) to observe what is going on around them and to press the gas pedal or the brake, as appropriate. The standard "follow-the-leader" theory supposes that

$$\ddot{u}_n(t+T) = \lambda\big(\dot{u}_{n-1}(t) - \dot{u}_n(t)\big), \qquad (\dagger)$$

where t is time, T is the reaction time, u_n is the position of the nth car, and the "sensitivity coefficient" λ may depend on $u_{n-1}(t) - u_n(t)$, the spacing between cars, and/or $\dot{u}_n(t)$, the speed of the nth car. The idea behind this equation is this. A driver will tend to decelerate if he is going faster than the car in front of him, or if he is close to the car in front of him, and will tend to accelerate if he is going slower than the car in front of him. In addition, a driver (especially in light traffic) may tend to speed up or slow down depending on whether he is going slower or faster (respectively) than a "reasonable" speed for the road (often, but not always, equal to the posted speed limit). Since our road is circular, in this equation u_0 is interpreted as u_N, where N is the total number of cars.

The simplest version of the model is the one in which the "sensitivity coefficient" λ is a (positive) constant. Then we have a homogeneous linear differential/difference equation with constant coefficients for the velocities $\dot{u}_n(t)$. Obviously, there is a "steady-state" solution when all the velocities are equal and constant (i.e., traffic is flowing at a uniform speed), but what we are interested in is the stability of the flow, or the question of what effect is produced by small differences in the velocities of the cars. The solution of (†) will be a superposition of exponential solutions of the form

$$u_n(t) = \exp(\alpha t) v_n,$$

where the v_n's and α are (complex) constants, and the system will be unstable if the velocities are unbounded, i.e., there are any solutions where the real part of α is positive. Using vector notation, we have

$$\dot{\mathbf{u}}(t) = \exp(\alpha t)\mathbf{v}, \qquad \ddot{\mathbf{u}}(t + T) = \alpha \exp(\alpha T) \exp(\alpha t)\mathbf{v}.$$

Substituting back in (†), we get the equation

$$\alpha \exp(\alpha T) \exp(\alpha t)\mathbf{v} = \lambda(S - I) \exp(\alpha t)\mathbf{v},$$

where

$$S = \begin{pmatrix} 0 & 0 & \cdots & 0 & 1 \\ 1 & 0 & \cdots & 0 & 0 \\ 0 & 1 & \cdots & 0 & 0 \\ \vdots & \vdots & \ddots & \vdots & \vdots \\ 0 & 0 & \cdots & 1 & 0 \end{pmatrix}$$

is the "shift" matrix that, when it multiplies a vector on the left, cyclically permutes the entries of the vector. We can cancel the $\exp(\alpha t)$ on each side to get

$$\alpha \exp(\alpha T)\mathbf{v} = \lambda(S - I)\mathbf{v},$$

or

$$\left(S - \left(1 + \frac{\alpha}{\lambda}\exp(\alpha T)\right)I\right)\mathbf{v} = 0, \qquad (**)$$

which says that v is an eigenvector for S with eigenvalue

$$1 + \frac{\alpha}{\lambda}\exp(\alpha T).$$

Since the eigenvalues of S are the Nth roots of unity, which are evenly spaced around the unit circle in the complex plane, and closely spaced together for large N, there is potential instability whenever

$$1 + \frac{\alpha}{\lambda}\exp(\alpha T)$$

has absolute value 1 for some α with positive real part; that is, whenever

$$\left(\frac{\alpha T}{\lambda T}\right)e^{\alpha T}$$

can be of the form $\exp(i\theta) - 1$ for some αT with positive real part. Whether instability occurs depends on the value of the product λT. We can see this by plotting values of $z\exp(z)$ for $z = \alpha T = iy$ a complex number on the critical line Re $z = 0$, and comparing with plots of $\lambda T(e^{i\theta} - 1)$ for various values of the parameter λT:

```
syms y; expand(i*y*(cos(y) + i*sin(y)))


ans =

y*cos(y)*i - y*sin(y)



ezplot(-y*sin(y), y*cos(y), [-2*pi, 2*pi]); hold on
theta = 0:0.05*pi:2*pi;
plot((1/2)*(cos(theta) - 1), (1/2)*sin(theta), '-');
plot(cos(theta) - 1, sin(theta), ':')
plot(2*(cos(theta) - 1), 2*sin(theta), '--');
title('iye^{iy} and circles \lambda T(e^{i\theta}-1)')
hold off
```

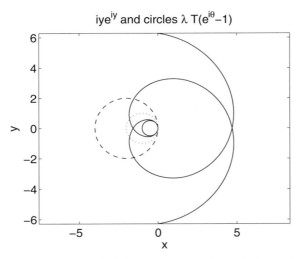

iye^{iy} and circles $\lambda\,T(e^{i\theta}-1)$

Here the small solid circle corresponds to $\lambda T = 1/2$, and we are just at the limit of stability, since this circle does not cross the spiral produced by $z\exp(z)$ for z a complex number on the critical line $\mathrm{Re}\,z = 0$, though it "hugs" the spiral closely. The dotted and dashed circles, corresponding to $\lambda T = 1$ or 2, do cross the spiral, so they correspond to unstable traffic flow.

We can check these theoretical predictions with a simulation using Simulink. We'll give a picture of the Simulink model and then explain it:

open_system traf

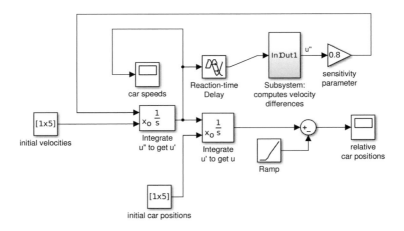

Here the Subsystem, which corresponds to multiplication by $S - I$, looks like this:

open_system 'traf/Subsystem: computes velocity differences'

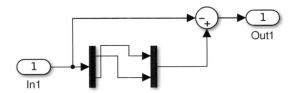

Most of the model is like the example in Chapter 8, except that our unknown function (called u), representing the car positions, is vector-valued, not scalar-valued. The major exceptions are these.

- We need to incorporate the reaction-time delay, so we've inserted a **Transport Delay** block from the **Continuous** block library, with the "Time delay" parameter T set to 0.5.
- The parameter λ shows up as the value of the gain in the "sensitivity parameter" **Gain** block in the upper right.
- Plotting car positions by themselves is not terribly useful, since only the relative positions matter. So before outputting the car positions to the **Scope** block labeled "relative car positions," we've subtracted a constant linear function (corresponding to uniform motion at the average car speed) created by the **Ramp** block from the **Sources** block library.
- We've made use of the option in the **Integrator** blocks to input the initial conditions, instead of having them built into the block. This makes the logical structure a little clearer.
- We've used the subsystem feature of Simulink. If you enclose a bunch of blocks with the mouse and then click on "Create subsystem" in the model's **Edit** menu, Simulink will package them as a subsystem. This is helpful if your model is large or if there is some combination of blocks that you expect to use more than once. Our subsystem sends a vector v to $(S - I)v = Sv - v$. A **Sum** block (with one of the signs changed to a minus) is used for vector subtraction. To model the action of S, we've used the **Demux** and **Mux** blocks from the **Signal Routing** block library. The **Demux** block, with the "number of outputs" parameter set to `[4, 1]`, splits a five-dimensional vector into a pair consisting of a four-dimensional vector and a scalar (corresponding to the last car). Then we reverse the order of these and put them back together with the **Mux** block, with the "number of inputs" parameter set to `[1, 4]`.

Once the model has been assembled, it can be run with various inputs. You can see the results yourself in the two **Scope** windows, but here we've run the simulation from the command line and plotted the results with the **simplot** command, that does almost the same thing as a **Scope** but in a regular MATLAB figure window. The following pictures are produced with $\lambda = 0.8$, corresponding to stable flow (though, to be honest, we've let two cars cross through each other briefly!):

```
set_param('traf/sensitivity parameter', 'Gain', '0.8');
trafout = sim('traf', 'SaveTime','on','TimeSaveName','t', ...
    'SaveState','on','StateSaveName','x');
t = trafout.find('t'); x = trafout.find('x');
```

The variable **t** stores the time parameter, the variable **x** stores car velocities in its first five columns and car positions in the second five columns. In this example, the average velocity is 3.15. First, we plot the relative positions, then we plot the velocities.

```
relpos = x(:,6:10) - 3.15*t*ones(1,5);
simplot(t, relpos), title('Relative Positions')
axis([0, 20, 0, 1]), axis normal
set(gcf, 'InvertHardcopy', 'off')
```

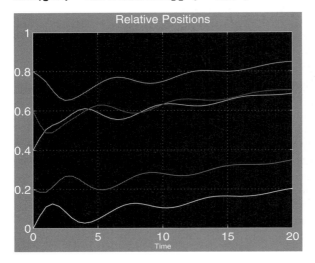

```
simplot(t, x(:,1:5)), title('Car Velocities')
axis([0, 20, 3, 3.3])
set(gcf, 'InvertHardcopy', 'off')
```

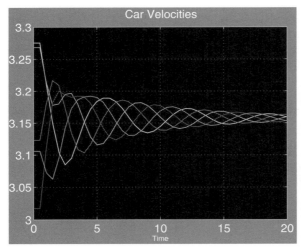

As you can see, the speeds fluctuate but eventually converge to a single value, and the separations between cars eventually stabilize. On the other hand, if λ is increased by changing the "sensitivity parameter" in the **Gain** block in the upper right, say from 0.8 to 2.0, we get the following output, which is typical of instability:

```
set_param('traf/sensitivity parameter', 'Gain', '2.0');
[t, x] = sim('traf');
relpos = x(:,6:10) - 3.15*t*ones(1,5);
simplot(t, relpos), title('Relative Positions')
axis([0, 20, -10, 10])
set(gcf, 'InvertHardcopy', 'off')
```

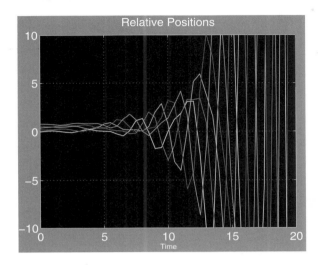

```
simplot(t, x(:,1:5)), title('Car Velocities')
axis([0, 20, -10, 10])
set(gcf, 'InvertHardcopy', 'off')
```

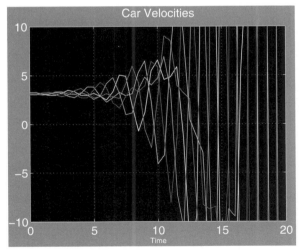

We encourage you to go back and tinker with the model (for instance using a sensitivity parameter that is also inversely proportional to the spacing between cars) and study the results.

Finally, you can create a movie with the following code:

```
clf reset
set(gcf, 'Color', 'White')
clear M
theta = (0:0.025:2)*pi;
for j = 1:length(t)
    plot(cos(x(j, 1:5)), sin(x(j, 1:5)), 'o');
    axis([-1, 1, -1, 1]);
    hold on; plot(cos(theta), sin(theta), 'r'); hold off
    axis equal;
    M(j) = getframe;
end
```

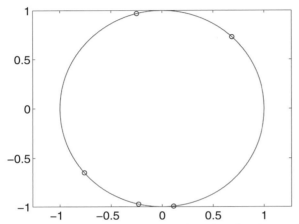

The idea here is that we have taken the circular road to have radius 1 (in suitable units), so that the command **plot(cos(theta),sin(theta),'r')** draws a red circle (representing the road) in each frame of the movie, and on top of that the cars are shown with moving little circles. The graph above is the last frame of the movie; you can view the entire movie by typing **movie(M)** or **movieview(M)**. Try it!

We should mention here one fine point needed to create a realistic movie. Namely, we need the values of **t** to be equally spaced – otherwise the cars will appear to be moving faster when the time steps are large, and will appear to be moving slower when the time steps are small. In its default mode of operation, Simulink uses a variable-step differential-equation solver based on MATLAB's command **ode45**, so the entries of **t** will not be equally spaced. We have fixed this by opening the **Configuration Parameters** dialog box using the **Simulation** menu in the model window, and, under the **Data Import/Export** item, changing the **Output options** box to read Produce specified output only, with **Output times** chosen to be **[0:0.5:20]**. Then the model will output the car positions only at times that are multiples of 0.5, and the MATLAB program above will produce a 41-frame movie.

Practice Set C

Developing Your MATLAB Skills

Problems 5–6 and 8–11 are a bit more advanced than the others. Problems 3 and 4 require either **linprog** from the Optimization Toolbox or else **simlp** from Simulink. Problems 6 and 10 require the Symbolic Math Toolbox; the others do not. Simulink is needed for Problems 7–9.

1. Consider an account that has M dollars in it and pays monthly interest J. Suppose that, beginning at a certain point, an amount S is deposited monthly and no withdrawals are made.

    (a) Assume first that $S = 0$. Using the *Mortgage Payments* application in Chapter 11 as a model, derive an equation relating J, M, the number n of months elapsed, and the total T in the account after n months. Assume that the interest is credited on the last day of the month and the total T is computed on the last day after the interest is credited.

    (b) Now assume that $M = 0$, that S is deposited on the first day of the month, and that as before interest is credited on the last day of the month, and the total T is computed on the last day after the interest has been credited. Once again, using the mortgage application as a model, derive an equation relating J, S, the number n of months elapsed, and the total T in the account after n months.

    (c) By combining the last two models, derive an equation relating all of M, S, J, n, and T, now of course assuming that there is an initial amount in the account (M) as well as a monthly deposit (S).

    (d) If the annual interest rate is 5%, and no monthly deposits are made, how many years does it take to double your initial stash of money? What if the annual interest rate is 10%?

    (e) In this and the next part, there is no initial stash. Assume an annual interest rate of 8%. How much do you have to deposit monthly to be a millionaire in 35 years (a career)?

    (f) If the interest rate remains as in (e) and you can afford to deposit only $300 each month, how long do you have to work to retire a millionaire?

    (g) You hit the lottery and win $100,000. You have two choices: take the money, pay the taxes, and invest what's left; or receive $100,000/240 monthly for 20 years, depositing what's left after taxes. Assume that a

207

$100,000 windfall costs you $35,000 in federal and state taxes, but that the smaller monthly payoff causes only a 20% tax liability. In which way are you better off 20 years later? Assume a 5% annual interest rate here.

(h) Historically, banks have paid roughly 5%, while the stock market has tended to return 8% on average over a ten-year period. (Bank rates have been much lower since the financial crisis of 2007–2008.) So parts (e) and (f) relate more to investing than to saving. But suppose that the market in a five-year period returns 13%, 15%, −3%, 5%, and 10% in five successive years, and then repeats the cycle. (Note that the [arithmetic] average is 8%, though a geometric mean would be more relevant here.) Assume that $50,000 is invested at the start of a five-year market period. How much does it grow to in five years? Now recompute four more times, assuming that you enter the cycle at the beginning of the second year, the third year, etc. Which choice yields the best/worst results? Can you explain why? Compare the results with a fixed-rate account paying 8%. Assume simple annual interest. Redo the five investment computations, assuming that $10,000 is invested at the start of each year. Again analyze the results.

2. Tony Gwynn, who played for the San Diego Padres 1982–2001, had a lifetime batting average of .338. This means that, for every 1000 at bats, he had 338 hits. (For this exercise, we shall ignore walks, hit batsmen, sacrifices, and other plate appearances that do not result in an official at bat.) Assume that in an average year he amassed 500 official at bats.

(a) Design a Monte Carlo simulation of a year in Tony's career. Run it. What is his batting average?

(b) Now simulate a 20-year career. Assume 500 official at bats every year. What is the best batting average in this "career"? What is the worst? What is the lifetime average?

(c) What do you think would happen if you ran the 20-year simulation 100 times and took the average of the lifetime averages for all 100 simulations?

3. Consider the following linear programming problem. Jane Doe is running for County Commissioner. She wants to personally canvass voters in the four main cities in the county: Gotham, Metropolis, Oz, and River City. She needs to figure out how many residences (private homes, apartments, etc.) to visit in each city. The constraints are as follows.

(i) She intends to leave a campaign pamphlet at each residence; she has only 50,000 available.

(ii) The travel costs she incurs for each residence are $0.50 in each of Gotham and Metropolis, $1 in Oz, and $2 in River City; she has $40,000 available.

(iii) The number of minutes (on average) that her visits to each residence require are 2 minutes in Gotham, 3 minutes in Metropolis, a minute in Oz, and 4 minutes in River City; she has 300 hours available.

(iv) Because of political profiles, Jane knows that she should not visit any more residences in Gotham than she does in Metropolis, and that, however many residences she visits in Metropolis and Oz, the total of the two should not exceed the number she visits in River City.

(v) Jane expects to receive, during her visits, on average, campaign contributions of one dollar from each residence in Gotham, a quarter from those in Metropolis, a half-dollar from the Oz residents, and three dollars from the folks in River City. She must raise at least $10,000 from her entire canvass.

Jane's goal is to maximize the number of supporters (those likely to vote for her). She estimates that for each residence she visits in Gotham the odds are 0.6 that she picks up a supporter, and the corresponding probabilities in Metropolis, Oz, and River City are, respectively, 0.6, 0.5, and 0.3.

(a) How many residences should she visit in each of the four cities?

(b) Suppose that she can double the time she can allot to visits. Now what is the profile for visits?

(c) But suppose that the extra time (in part (b)) also mandates that she double the contributions she receives. What is the profile now?

4. Consider the following linear programming problem. The famous football coach Joe Glibb is trying to decide how many hours to spend with each component of his offensive unit during the coming week – that is, the quarterback, the running backs, the receivers, and the linemen. The constraints are as follows:

(i) The number of hours available to Joe during the week is 50.

(ii) Joe figures he needs 20 points to win the next game. He estimates that for each hour he spends with the quarterback, he can expect a point return of 0.5. The corresponding numbers for the running backs, receivers, and linemen are 0.3, 0.4, and 0.1, respectively.

(iii) In spite of their enormous size, the players are relatively thin skinned. Each hour with the quarterback is likely to require Joe to criticize him once. The corresponding numbers of criticisms per hour for the other three groups are 2 for running backs, 3 for receivers, and 0.5 for linemen. Joe figures he can bleat out only 75 criticisms in a week before he loses control.

(iv) Finally, the players are *prima donnas* who engage in rivalries. Because of that, he must spend exactly the same number of hours with the running backs as he does with the receivers, at least as many hours with the quarterback as he does with the runners and receivers combined, and at least as many hours with the receivers as with the linemen.

Joe worries that he's going to be fired at the end of the season regardless of the outcome of the game, so his primary goal is to maximize his pleasure during the week. (The team's owner should only know.) He estimates that, on a sliding scale from 0 to 1, he gets 0.2 units of personal satisfaction for each hour with the quarterback. The corresponding numbers for the runners, receivers, and linemen are 0.4, 0.3, and 0.6, respectively.

(a) How many hours should Joe spend with each group?

(b) Suppose that he needs only 15 points to win; then how many?

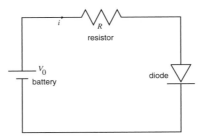

Figure C.1. A Nonlinear Circuit.

5. This problem, suggested to us by our colleague Tom Antonsen, concerns an electrical circuit, one of whose components does not behave linearly. Consider the circuit in Figure C.1. Unlike the resistor, the diode is a non-linear element – it does not obey Ohm's Law. In fact its behavior is specified by the formula

$$i = I_0 \exp(V_D/V_T), \qquad\qquad (C.1)$$

where i is the current in the diode (which is the same as that in the resistor by Kirchhoff's Current Law), V_D is the voltage across the diode, I_0 is the leakage current of the diode, and $V_T = kT/e$, where k is Boltzmann's constant, T is the temperature of the diode, and e is the electrical charge.

Now, by application of Ohm's Law to the resistor, we also know that $V_R = iR$, where V_R is the voltage across the resistor and R is its resistance. But, by Kirchhoff's Voltage Law, we also have $V_R = V_0 - V_D$. This gives a second equation relating the diode current and voltage, namely

$$i = (V_0 - V_D)/R. \qquad\qquad (C.2)$$

Note now that (C.2) says that i is a decreasing linear function of V_D with value V_0/R when V_D is zero. At the same time, (C.1) says that i is an exponentially growing function of V_D starting out at I_0. Since, typically, $RI_0 < V_0$, the two resulting curves (for i as a function of V_D) must cross once. Eliminating i from the two equations, we see that the voltage in the diode must satisfy the transcendental equation

$$(V_0 - V_D)/R = I_0 \exp(V_D/V_T),$$

or

$$V_D = V_0 - RI_0 \exp(V_D/V_T).$$

(a) Reasonable values for the electrical constants are $V_0 = 1.5$ volts, $R = 1000$ ohms, $I_0 = 10^{-5}$ amperes, and $V_T = 0.0025$ volts. Use **fzero** to find the voltage V_D and current i in the circuit.

(b) In the remainder of the problem, we assume that the voltage V_0 in the battery and the resistance R of the resistor are unchanged. But suppose that we have some freedom to alter the electrical characteristics of the diode. For example, suppose that I_0 is halved. What happens to the voltage?

(c) Suppose that, instead of halving I_0, we halve V_T. Then what is the effect on V_D?

(d) Suppose that both I_0 and V_T are cut in half. What then?

(e) Finally, we want to examine the behavior of the voltage if both I_0 and V_T are decreased toward zero. For definitiveness, assume that we set $I_0 = 10^{-5}x$ and $V_T = 0.0025x$, and let $x \rightarrow 0$. Specifically, compute the solution for $x = 10^{-j}$, $j = 0, \ldots, 5$. Then, display a **loglog** plot of the solution values, for the voltage as a function of I_0. What do you conclude?

6. This problem is based on the *Population Dynamics* and the *360° Pendulum* applications from Chapter 11. The growth of a species was modeled in the former by a *difference equation*. In this problem, we will model population growth by a *differential equation*, akin to the second application mentioned above. In fact, we can give a differential-equation model for the logistic growth of a population x as a function of time t by means of the equation

$$\dot{x} = x(1 - x) = x - x^2, \qquad \text{(C.3)}$$

where $\dot{x}$ denotes the derivative of x with respect to t. We think of x as a fraction of some maximal possible population. One advantage of this continuous model over the discrete model in Chapter 11 is that we can get a "reading" of the population at any point in time (not just on integer intervals).

(a)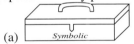

   The differential equation (C.3) is solved in any beginning course in ordinary differential equations, but you can do it easily with the MATLAB command **dsolve**. (Look up the syntax via online help.)

(b) Now find the solution assuming an initial value $x_0 = x(0)$ of x. Use the values $x_0 = 0, 0.25, \ldots, 2.0$. Graph the solutions and use your picture to justify the statement: "Regardless of $x_0 > 0$, the solution of (C.3) tends to the constant solution $x(t) \equiv 1$ in the long term."

The logistic model presumes two underlying features of population growth: (i) that ideally the population expands at a rate proportional to its current total (i.e., exponential growth – this corresponds to the x term on the right-hand side of (C.3)); and (ii) because of interactions between members of the species and natural limits to growth, unfettered exponential growth is held in check by the logistic term, given by the $-x^2$ expression in (C.3). Now assume that there are two species $x(t)$ and $y(t)$, competing for the same resources to survive. Then there will be another negative term in the differential equation that reflects the interaction between the species. The usual model presumes it to be proportional to the product of the two populations, and the larger the constant of proportionality, the more severe the interaction, as well as the resulting check on population growth.

(c) Here is a typical pair of differential equations that model the growth in population of two competing species $x(t)$ and $y(t)$:

$$\dot{x}(t) = x - x^2 - 0.5xy$$
$$\dot{y}(t) = y - y^2 - 0.5xy. \tag{C.4}$$

The command **dsolve** can solve many pairs of ordinary differential equations – especially linear ones. But the mixture of quadratic terms in (C.4) makes it unsolvable symbolically, so we need to use a numerical ODE solver as we did in the pendulum application. Using the commands in that application as a template, graph numerical solution curves to the system (C.4) for initial conditions $x(0) = j/12$, $y(0) = k/12$, $0 \leq j, k \leq 13$. (*Hint*: use **axis** to limit your view to the square $0 \leq x, y \leq 13/12$.)

(d) The picture you drew is called a *phase portrait* of the system. Interpret it. Explain the long-term behavior of any population distribution that starts with only one species present. Relate it to part (b). What happens in the long term if both populations are present initially? Is there an initial population distribution that remains undisturbed? What is it? Relate those numbers to the model (C.4).

(e) Now replace 0.5 in the model by 2, that is consider the new model

$$\dot{x}(t) = x - x^2 - 2xy$$
$$\dot{y}(t) = y - y^2 - 2xy. \tag{C.5}$$

Draw the phase portrait. (Use the same initial data and viewing square.) Answer the same questions as in part (d). Do you see a special solution trajectory that emanates from near the origin and proceeds to the special fixed point? Do you see another trajectory from the upper right to the fixed point? What happens to all population distributions that do not start on these trajectories?

(f) Explain why model (C.4) is called "peaceful coexistence" and model (C.5) is called "doomsday." Now explain heuristically why the change in coefficient from 0.5 to 2 converts coexistence into doomsday.

7.   Build a Simulink model corresponding to the pendulum equation

$$\ddot{x}(t) = -0.5\dot{x}(t) - 9.81\sin(x(t)) \tag{C.6}$$

from the *360° Pendulum* in Chapter 11. You will need the Trigonometric Function block from the Math Operations library. Use your model to redraw some of the phase portraits.

8. ◤◣ As you know, Galileo and Newton discovered that all bodies near the Earth's surface fall with the same acceleration g due to gravity, approximately 32.2 ft/sec^2. However, real bodies are also subjected to forces due to air resistance. If we take both gravity and air resistance into account, a moving baseball can be modeled by the differential equation

$$\ddot{\mathbf{x}} = [0, -g] - c\,\|\dot{\mathbf{x}}\|\,\dot{\mathbf{x}}. \tag{C.7}$$

Here $\mathbf{x}$, a function of the time t, is the vector giving the position of the ball (the first coordinate is measured horizontally, the second one vertically), $\dot{\mathbf{x}}$ is the velocity vector of the ball, $\ddot{\mathbf{x}}$ is the acceleration of the ball, $\|\dot{\mathbf{x}}\|$ is the magnitude of the velocity, that is, the speed, and c is a constant depending on the surface characteristics and mass of the ball and the density of the air. (We are neglecting the lift force that comes from the ball's rotation, which can also play a major role in some situations, for instance in analyzing the path of a curve ball, as well as forces due to wind currents.) For a baseball, the constant c turns out to be approximately 0.0017, assuming that distances are measured in feet and time is measured in seconds. (See, for example, Chapter 18, "Balls and Strikes and Home Runs," in *Towing Icebergs, Falling Dominoes, and Other Adventures in Applied Mathematics*, by Robert Banks, Princeton University Press, Princeton, 1998.) Build a Simulink model corresponding to equation (C.7), and use it to study the trajectory of a batted baseball. Here are a few hints. Represent $\ddot{\mathbf{x}}$, $\dot{\mathbf{x}}$, and $\mathbf{x}$ as vector signals, joined by two Integrator blocks. The quantity $\ddot{\mathbf{x}}$, according to (C.7), should be computed from a Sum block with two vector inputs. One should be a Constant block with the vector value $[0, -32.2]$, representing gravity, and the other should represent the drag term on the right-hand side of equation (C.7), computed from the value of $\dot{\mathbf{x}}$. You should be able to change one of the parameters to study what happens both with and without air resistance (the cases of $c = 0.0017$ and $c = 0$, respectively). Attach the output to an XY Graph block, with the parameters x-min = 0, y-min = 0, x-max = 500, y-max = 150, so you can see the path of the ball out to a distance of 500 feet from home plate and up to a height of 150 feet.

(a) Let $\mathbf{x}(0) = [0, 4]$ and $\dot{\mathbf{x}}(0) = [80, 80]$. (This corresponds to the ball starting at $t = 0$ from home plate, 4 feet off the ground, with the horizontal and vertical components of its velocity both equal to 80 ft/sec. This corresponds to a speed off the bat of about 77 mph, which is not unrealistic.) How far (approximately – you can read this from your XY Graph output) will the ball travel before it hits the ground, both with and without air resistance? About how long will it take the ball to hit the ground, and how fast will the ball be traveling at that time (again, both with and without air resistance)? (The last parts of the question are relevant for outfielders.)

(b) Suppose that a game is played in Denver, Colorado, where because of thinning of the atmosphere due to the high altitude, c is only 0.0014. How far will the ball travel now (given the same initial velocity as in (a))?

(c) (This is not a MATLAB problem.) Estimate from a comparison of your answers to (a) and (b) what effect altitude might have on the team batting average and team earned run average of the Colorado Rockies.

9. ▶️■ The Black–Scholes model, mentioned in the *Financial Data* application, can be reformulated in terms of a stochastic differential equation, that is a differential equation with random noise in it:

$$\frac{1}{S}\frac{dS}{dt} = C + \text{random noise}.$$

Here S is a stock price or stock index price, C is a constant (which if there were no noise would give exponential growth or decay for S, and t is time. Construct a Simulink model for this equation, using the Band-Limited White Noise block for the noise term and a Constant block for C. You can use the initial condition $S(0) = 1$ (in suitable units). Run the model a few times with different values of C and of the "Noise power" parameter of the noise block, and see how the graph of S as a function of t (displayed in a Scope block) compares with, say, the graph of the Dow Jones average in the *Financial Data* application. To start off, try taking $C = 0.1$, the "Noise Power" equal to 0.2, and the "Sample time" parameter of the noise block equal to 0.01.

10. ∫ƒ An *elliptic curve* is a curve E of the form $y^2 = x^3 + ax + b$, where a and b are constants. Traditionally, we allow a special point ∞ on this curve as well. There is an "addition operation" on E, for which ∞ acts as the zero element, defined by the requirements that $-(x, y) = (x, -y)$ and that $P + Q + R = \infty$ if P, Q, and R are distinct collinear points on E. We also define $2P$ by $2P + R = \infty$, where R is the other point on E at which the tangent line to E at P intersects the curve. Write a MuPAD procedure to compute "doubling" on E and compute $P, 2P, 4P, 8P, \cdots, 2^6 P$ when E is the curve $y^2 = x^3 + x + 6$ and P is the point $(-1, 2)$ (which clearly lies on E). What happens to the coordinates of the points as you get far out in the sequence?

11. ☆ Consider a sequence of complex numbers generated from a starting value z_0 by the rule $z_{n+1} = z_n^2 - 0.75$. For some values of z_0, the sequence of numbers will diverge to infinity as n increases, but for other values of z_0, the sequence will remain inside a bounded region forever. The boundary of the set consisting of the latter values of z_0 is called the *Julia set* for the function $f(z) = z^2 - 0.75$. Use **image** or **imagesc** to draw a picture of this Julia set. (*Hint*: the Julia set lies within the region where the real part of z_0 is between -2 and 2, and its imaginary part is between -1.5 and 1.5. Form a grid of z_0 values within this rectangle (we suggest using roughly a 300×400 array) and calculate the corresponding values of $z_1, z_2, \ldots, z_n$ for some n. Color the pixel of each z_0 according to how large the corresponding z_n is. The larger you make n, the more accurate (and interesting) the picture will be, but the computation will also take longer, so start with a relatively small value of n and work your way up.)

Chapter 12

Troubleshooting

In this chapter, we offer advice for dealing with some common problems that you may encounter. We also list and describe the most common mistakes that MATLAB users make. Finally, we offer some simple but useful techniques for debugging your M-files.

Common Problems

Problems manifest themselves in various ways: totally unexpected or plainly wrong output appears; MATLAB produces an error message (or at least a warning); MATLAB refuses to process an input line; something that worked earlier stops working; or, worst of all, the computer freezes. Fortunately, these problems are often caused by several easily identifiable and correctable mistakes. What follows is a description of some common problems, together with a presentation of likely causes, suggested solutions, and illustrative examples. We also refer to places in the book where related issues are discussed. Here is a list of the problems:

- Wrong or unexpected output

- Syntax error

- Undefined function

- Error or warning messages when plotting

- A previously saved M-file evaluates differently

- Computer won't respond.

⇨ **As we mentioned in Chapter 6, warning messages can be suppressed, but for troubleshooting it is best to make sure they are displayed. Type `warning` to see whether warning messages are turned on; you can enable display of all warning messages with `warning on all`.**

Wrong or Unexpected Output

There are many possible causes for this problem, but they are likely to be among the following:

CAUSE: **Forgetting old variable definitions.**

SOLUTION: Be careful not to use variables you don't remember defining, and clear variables you no longer need, especially in a long session.

215

☞   *See* Naming and Clearing Variables *in Chapter 2.*

EXAMPLE: Suppose that you set up a loop such as

```
>> for i = 1:10
...
end
```

and later in your session you enter the complex number **2+i**; then in fact you have entered the number 12, because at the end of the loop the variable **i** is set equal to 10. In order to restore it to its intrinsic value $\sqrt{-1}$, you need to type **clear i**.

CAUSE: **Conflicting definitions.**

SOLUTION: Do not use the same name for two different functions or variables, and, in particular, do not overwrite the names of any of MATLAB's built-in functions.

You can accidentally mask one of MATLAB's built-in M-files either with your own M-file of the same name or with a variable (including, perhaps, an anonymous function). When unexpected output occurs and you think this might be the cause, it helps to use **which** to find out what M-file, function, or variable is actually being used. The preceding example illustrates this problem, but here is perhaps a more extreme example.

EXAMPLE:

```
>> plot = gcf;
>> X = -2:0.1:2;
>> plot(X, X.^2)
```

```
Subscript indices must either be real positive integers or
logicals.
```

What's wrong, of course, is that **plot** has been masked by a variable with the same name. You could detect this with:

```
>> which plot
```

```
plot is a variable.
```

If you type **clear plot** and execute the **plot** command again, you'll get a picture of the desired parabola. A more subtle example could occur if you redefined **plot** on purpose, not thinking you would use the built-in command, and then ran a script M-file that uses **plot** internally.

CAUSE: **Attempting to refer to old output with ans.**

SOLUTION: Assign variable names to any output that you intend to use.

If you decide at some point in a session that you wish to refer to prior output that was unnamed, then execute the command again with the output assigned to a new variable. (The UP-ARROW key or Command History Window is useful for recalling the command to edit it.) If you rely on **ans**, even when it works as intended, you have less flexibility to edit your input – for example, to fix mistyped input in the Command Window, or to insert commands into an M-file.

CAUSE: **Improper use of built-in functions.**

SOLUTION: Always use the names of built-in functions exactly as MATLAB specifies them; always enclose inputs in parentheses, not brackets and not braces; always list the inputs in the required order.

☞   *See* Managing Variables *and* Online Help *in Chapter 2.*

CAUSE: **Inattention to precedence of arithmetic operations.**

SOLUTION: Use parentheses liberally and correctly when entering arithmetic or algebraic expressions.

EXAMPLE: MATLAB first exponentiates, then divides and multiplies, and finally adds and subtracts, unless a different order is specified by using parentheses. So if you attempt to compute $9^{1/2} - (4/2)^2$ by typing

```
>> 9^1/2 - 4/2^2
ans =
    3.5000
```

the answer MATLAB produces is not what you intended because **9** is raised to the power **1** before the division by **2**, and **2** is squared before being divided into **4**. Here is the correct calculation:

```
>> 9^(1/2) - (4/2)^2
ans =
    -1
```

Syntax Error

CAUSE: **Mismatched delimiters (parentheses, quote marks, braces, or brackets).**

SOLUTION: Look carefully at the input line to find a missing or an extra delimiter.

If you make this mistake in the Command Window, the resulting error message points to the location in the command line where MATLAB first detected the problem; this is often the place where you need to add or remove a delimiter. The Editor/Debugger highlights questionable syntax with a wavy red underline, though in some cases this is a false alarm; move the mouse pointer over the highlighted location to see what MATLAB thinks the problem is. In addition, in both the Command Window and the Editor/Debugger, MATLAB briefly highlights matching delimiters while you type, and color-codes strings (expressions enclosed in single quotes) so that you can see where they begin and end.

☞   *If you never see the wavy red underlining in the Editor/Debugger, perhaps you have the MATLAB Code Analyzer turned off; see Debugging Techniques below.*

CAUSE: **Wrong delimiters: using parentheses in place of brackets, or vice versa, and so on.**

SOLUTION: Remember the basic rules about delimiters in MATLAB.

Parentheses are used both for grouping arithmetic expressions and for enclosing inputs to a MATLAB command, an M-file, or an anonymous function. They are also used for referring to an entry in a vector or matrix. *Square brackets* are used for defining vectors or matrices. *Single quote marks* are used for defining strings.

EXAMPLE: The following illustrates what can happen if you don't follow these rules:

```
>> X = -1:0.01:1;
>> X[1]

 X[1]
   |
Error: Unbalanced or unexpected parenthesis or bracket.


>> A = (0,1,2)

 A = (0,1,2)
       |
Error: Expression or statement is incorrect--possibly
unbalanced (, {, or [.
```

In the first case, **X(1)** was intended, and in the second case, **A = [0,1,2]** was intended. But the second error message is harder to understand; it points at the first comma because it expected the parentheses to be closed before the comma. Here is a more subtle example:

```
>> abs -3

ans =
    45      51
```

This time there's no error message, but the answer is certainly not the absolute value of -3. Recall that when a MATLAB command is separated from its input by only a space (as in **clear x**), the input is interpreted as a string. Thus, MATLAB has interpreted **-3** not as the number -3, but as the string **'-3'**! And sure enough, the hyphen and the character "3" convert to the numbers 45 and 51 in the usual numerical encoding of strings.

```
>> double('-3')

ans =
45      51
```

Of course, to get the absolute value of -3, you should type **abs(-3)**.

Braces or *curly brackets* are used less often than either parentheses or square brackets, and are usually not needed by beginners. Their main use is with cell arrays. One example to keep in mind is that if you want an M-file to take a variable number of inputs or produce a variable number of outputs, then these are stored in the cell arrays **varargin** and **varargout**, and braces are used to refer to the cells of these arrays. Similarly, **case** is sometimes used with braces in the middle of a **switch** construct. If you want to construct an array of strings, then it has to be done with braces, since brackets when applied to strings are interpreted as concatenation.

EXAMPLE:

```
>> {'a', 'b'}
```

```
ans =
    'a'      'b'
```

`>> ['a', 'b']`

```
ans =
ab
```

CAUSE: **Improper use of arithmetic symbols.**

SOLUTION: When you encounter a syntax error, review your input line carefully for mistakes in typing.

EXAMPLE: If the user, intending to compute 2 times −4, inadvertently switches the symbols, the result is:

`>> 2-*4`

```
 2-*4
    |
Error: Unexpected MATLAB operator.
```

Here the vertical bar highlights the place where MATLAB believes the error is located. In this case, the actual error is earlier in the input line.

Undefined Function

CAUSE: **Using uppercase instead of lowercase letters in MATLAB commands, misspelling a command, or misremembering its name.**

SOLUTION: Fix the spelling, or consult the online help for the correct name.

In the Command Window, MATLAB will often suggest a correction for a mildly mistyped command.

EXAMPLE:

`>> arctan(1)`

```
Undefined function 'arctan' for input arguments of type
'double'.
```

```
Did you mean:
```

`>> atan(1)`

```
ans =
    0.7854
```

In the example above, MATLAB displayed the corrected input line and put the cursor at the end of the line, giving the option of editing the line or just executing it (which we did).

CAUSE: **Attempting to run an M-file that is not in your current folder or search path**

SOLUTION: Change your current folder, or add the folder that contains the M-file to your search path.

☞   *See* The Current Folder and Search Path *in Chapter 3.*

If you run an M-file from the Editor/Debugger, you will be prompted to change your current folder or search path if necessary.

Error or Warning Messages When Plotting

CAUSE: **There are several possible explanations, but often the problem is the wrong type of input for the plotting command chosen, or graphing a function that is undefined at some values.**

SOLUTION: Carefully follow the examples in the online help for the plotting command, and pay attention to the error and warning messages.

EXAMPLE:

```
>> X = -3:0.01:3;
>> plot(X, X.^(1/3))
```

```
Warning: imaginary parts of complex X and/or Y arguments
ignored.
```

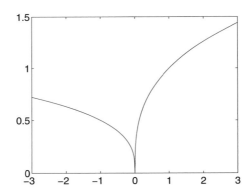

Figure 12.1. A Spurious Graph of $y = x^{1/3}$.

Figure 12.1 is not the graph of $y = x^{1/3}$ on the interval $-3 \leq x \leq 3$. Actually, the graph to the right of the origin looks correct, but to the left of the origin the graph is certainly not right. The warning message provides a clue: **plot** is intended to graph real numbers, but somehow we have supplied it with complex numbers. In that case, MATLAB plots the real part of the input. It turns out that for negative values of x, MATLAB computed $x^{1/3}$ to be the complex cube root $[(1 + \sqrt{3}i)/2](-x)^{1/3}$. In order to fix the problem, we must specify the function in MATLAB more carefully. The correct graph in Figure 12.2 results from:

```
>> plot(X, sign(X).*abs(X).^(1/3))
```

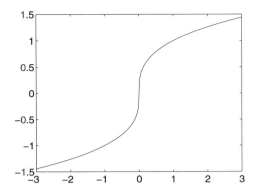

Figure 12.2. Correct Graph of $y = x^{1/3}$.

A Previously Saved M-File Evaluates Differently

One of the most frustrating problems you may encounter occurs when a previously saved script M-file, which worked as intended when you saved it, won't evaluate or evaluates incorrectly when opened in a new session.

CAUSE: **Usually, failure to initialize variables.**

SOLUTION: Make sure to define initial values for all variables that you use in a script M-file; if you want your M-file to depend on input defined in the Command Window, use a function M-file.

Your previous successful run of the M-file may have relied upon variables that you defined in the Command Window. If you haven't defined these variables in the new session, you'll get an error message when you run the M-file. If you defined variables used in the M-file differently in the new session, you'll get different output. Clearing variables at the beginning of the M-file helps to detect uninitialized variables.

EXAMPLE: The problem is illustrated by the following simple, but poorly designed, program. To compute 10! (10-factorial), a user writes the following script M-file:

```
%% computing 10!
for k = 1:10
    f = f*k;
end
f
```

The user may have, upon encountering an error, defined **f=1** in the Command Window and then obtained the correct answer by running the M-file. But having failed to fix the M-file by adding **f=1;** at the beginning, the user is doomed to debug the problem again when using the M-file in a new session, or get the wrong answer due to inadvertently defining a different value for **f**.

Computer Won't Respond

CAUSE: **MATLAB is caught in a very long calculation, or some other calamity has caused it to fail to respond. Perhaps you have defined, or a MATLAB command has created, an array that is too large for your computer memory to handle.**

SOLUTION: Abort the calculation by typing CTRL+C in the Command Window.

If overuse of computer memory is the problem, try to redo your calculation using smaller arrays, e.g., by using fewer grid points in a 3D plot, or by breaking a large vectorized calculation into smaller pieces using a loop. Clearing large arrays from your Workspace may help too.

EXAMPLE: You'll know it when you see it!

The Most Common Mistakes

The most common mistakes are all accounted for in the causes of the problems described earlier. But to help you prevent these mistakes, we compile them here in a single list to which you can refer periodically. Doing so will help you to establish "good MATLAB habits."

- Forgetting to clear or initialize variables

- Improper use of built-in functions

- Inattention to the order of precedence of arithmetic operations

- Mismatched delimiters

- Using the wrong delimiters

- Plotting the wrong kind of object

Debugging Techniques

Now that we have discussed the most common mistakes, it's time to discuss how to debug your M-files, and how to locate and fix those pesky problems that don't fit into the neat categories above.

If one of your M-files is not working the way you expected, perhaps the easiest thing you can do to debug it is to insert the command **keyboard** somewhere in the middle. This temporarily suspends (but does not stop) execution and returns command to the keyboard, where you are given a special prompt with a **K** in it. You can execute whatever commands you want at this point (for instance, to examine some of the variables). To return to execution of the M-file, type **return** or **dbcont**, short for "debug continue."

A more systematic way to debug M-files is to use the debugging features of the MATLAB M-file Editor/Debugger. Start by checking warning messages from the Code Analyzer. You can see these messages by moving the mouse pointer over the horizontal yellow lines on the right edge of the Editor/Debugger window, or on a wavy red underline within the window. Click on a yellow line to move the cursor to the location referred to by the corresponding warning message. If there are a lot of yellow lines, you may prefer to view all the warning messages simultaneously in a separate window as follows. In the upper right corner of the Editor/Debugger window, click on the small circle with an inverted triangle inside, then select "Show Code Analyzer Report". Alternatively, you can use **checkcode** in the Command Window.

✓ To turn the Code Analyzer off or on, go to the Desktop window and select HOME:ENVIRONMENT:**Preferences**, then select "Code Analyzer" in the window that appears. In addition to turning all warnings off or on, you can turn off specific types of warnings. For example, if you are preparing a script M-file to publish, you probably want many of your commands to produce output; thus, you may want to turn off the message "Terminate statement with semicolon to suppress output (within a script)." You can also turn off messages for a particular line, a particular file, or all files from the Editor/Debugger by right-clicking on the underlined location that produced a warning and selecting one of the options from the line that starts "Suppress".

To diagnose problems further, you can insert "breakpoints" in the M-file. In the Editor/Debugger, you can set a breakpoint on any line with a dash to the right of its line number, by clicking on the dash. You can also set breakpoints via the EDITOR:BREAKPOINTS:**Breakpoints** menu, or from the command line with **dbstop**. Once a breakpoint has been inserted in the M-file, the dash to the right of the line number changes to a red dot (an example is illustrated in Figure 12.8 below.) Then, when you run the M-file, execution will pause just before the line with the breakpoint. A green arrow will appear next to the red circle and, as with **keyboard**, a special prompt with a **K** in it will appear in the Command Window. At the prompt, you can check the values of variables defined in the M-file, and when you are ready to continue you have several options, which appear in the right side of the EDITOR tab in the Editor/Debugger. The two simplest options are **Step**, which executes the line with the green arrow and pauses at the next executable line, and **Continue**, which continues execution until the next breakpoint, if any. You can perform the same operations from the Command Window with **dbstep** and **dbcont**. If you continue from a breakpoint in a loop, execution will pause at the same breakpoint the next pass through the loop. To clear a breakpoint, click on the red circle, or to clear all breakpoints use the EDITOR:BREAKPOINTS:**Breakpoints** menu (you can also use **dbclear**).

✓ There are many more options, including "conditional breakpoints," which you can set by right-clicking on the red circle of an existing breakpoint. For example, if you have a line containing **log(x)** in a loop, and at some unknown iteration of the loop you get an error because **x** is negative, you can set a breakpoint to pause execution at this line only when **x** is negative. Another option is to pause execution whenever a warning message is generated; use the EDITOR:BREAKPOINTS:**Breakpoints** menu or the command **dbstop if warning**. See the topic "Debugging Process and Features" in the online help for a more detailed overview.

Let's illustrate these techniques with a real example. Suppose that you want to construct a function M-file that takes as input two functions f and g and two numbers

a and b, plots f and g between $x = a$ and $x = b$, and shades the region in between them. As a first try, you might start with the function M-file shadecurves.m given as follows:

```
function shadecurves(f, g, a, b)
%SHADECURVES   Draws the region between two curves
%    SHADECURVES(f, g, a, b) inputs two functions f
%    and g, plots them between x = a and x = b, and
%    shades the region in between.
%    Example: shadecurves(@sin, @(x) -sin(x), 0, pi)
fv = str2func(vectorize(func2str(f)));
gv = str2func(vectorize(func2str(g)));
xvals = linspace(a, b, 100);
plot([xvals xvals], [fv(xvals) gv(xvals)])
```

Let's run the M-file with the data specified in the help lines:

```
>> shadecurves(@sin, @(x) -sin(x), 0, pi)
```

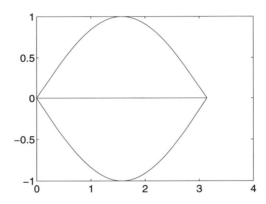

Figure 12.3. A First Attempt at Shading the Region between $\sin x$ and $-\sin x$.

The graph in Figure 12.3 is not really what we wanted; the figure we seek is shown in Figure 12.4. On the other hand, we do seem to have successfully graphed the two input functions on the desired interval. Let's try another example to test that the M-file is able to handle an input function that is not vectorized:

```
>> shadecurves(@(x) x^2, @sqrt, 0, 1)
```

The resulting Figure 12.5 shows the graphs of the two input functions, though again it shows an extra line as well. Let's try to fix the lack of shading first. Of course, we haven't put anything in the **plot** command to specify shading, and the online help doesn't indicate any shading options. Searching the online help for the word "shaded" leads to the command **fill** for 2-D "filled" plots. That suggests changing the last line of the M-file to:

```
fill([xvals xvals], [fv(xvals) gv(xvals)])
```

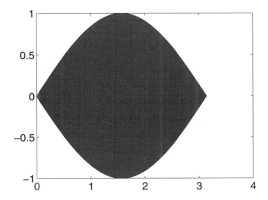

Figure 12.4. The Shaded Region between $\sin x$ and $-\sin x$.

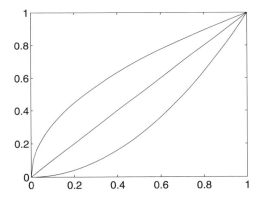

Figure 12.5. A First Attempt at Shading the Region between x^2 and $\sqrt{x}$.

But running the M-file again yields the error message:

```
Error using fill
Not enough input arguments.

Error in shadecurves (line 10)
fill([xvals xvals], [fv(xvals) gv(xvals)])
```

Checking the online help for **fill** more carefully, we see that it requires a third input, which can (among other options) be a fill color in RGB coordinates. So we change our final line to, for instance:

fill([xvals xvals], [ffun(xvals) gfun(xvals)], [.8 0 1])

Repeating the command **shadecurves(@(x) x^2, @sqrt, 0, 1)** yields the picture shown in Figure 12.6.

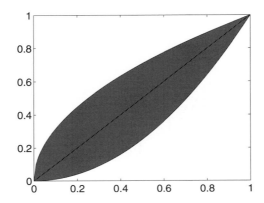

Figure 12.6. A Second Attempt at Shading the Region between x^2 and $\sqrt{x}$.

The result is better, but the diagonal line is not covered up by the shading. Not only that, Figure 12.7 shows another example with two spurious lines that affect the filled region:

```
>> shadecurves(@(x) x^2, @(x) x^4, -1.5, 1.5)
```

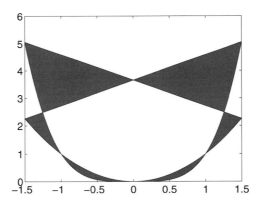

Figure 12.7. A First Attempt at Shading the Region between x^2 and x^4.

To understand what is happening, we should examine carefully the input to **fill** and determine what **fill** is designed to do with that input. The online help indicates that **fill** draws a filled two-dimensional polygon whose vertices are given by the horizontal coordinates in the first input vector and the vertical coordinates in the second input vector. The input is not accessible after the M-file is run, because variables in a function M-file are not saved to the Workspace. So, we'll use a breakpoint before the last line (line 10) of the M-file to examine the input before **fill** is executed. To simplify the input, we also change **100** to **5** on line 9. At this point, our M-file in the Editor/Debugger window now looks like Figure 12.8. Note the circle on line 10, indicating the breakpoint. We repeat the command **shadecurves(@(x) x^2, @(x) x^4, -1.5, 1.5)** and go the the Command Window to check the input to **fill**, converting the first two inputs to columns of a matrix:

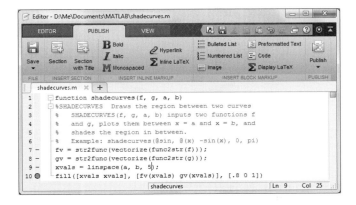

Figure 12.8. The Editor/Debugger.

```
K>> [[xvals xvals]', [fv(xvals) gv(xvals)]']

ans =
   -1.5000      2.2500
   -0.7500      0.5625
         0           0
    0.7500      0.5625
    1.5000      2.2500
   -1.5000      5.0625
   -0.7500      0.3164
         0           0
    0.7500      0.3164
    1.5000      5.0625
```

Typing

```
K>> dbcont
```

then yields Figure 12.9.

Now we can follow on the graph the 10 points, whose coordinates we just printed, that MATLAB has connected into a polygon. One of the unintended lines occurred when MATLAB connected the point $(1.5, 2.25)$ to the point $(-1.5, 5.0625)$. This is not what we wanted; we wanted MATLAB to join the point $(1.5, 2.25)$ on the curve $y = x^2$ to the point $(1.5, 5.0625)$ on the curve $y = x^4$. We can fix this by reversing the order of the x-coordinates at which we evaluate the second function g. So, letting **slavx** denote **xvals** reversed, via **fliplr**, we correct our M-file to read:

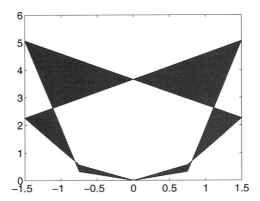

Figure 12.9. A Second Attempt at Shading the Region between x^2 and x^4.

```
function shadecurves(f, g, a, b)
%SHADECURVES  Draws the region between two curves
%    SHADECURVES(f, g, a, b) inputs two functions f
%    and g, plots them between x = a and x = b, and
%    shades the region in between.
%    Example: shadecurves(@sin, @(x) -sin(x), 0, pi)
fv = str2func(vectorize(func2str(f)));
gv = str2func(vectorize(func2str(g)));
xvals = linspace(a, b, 100); slavx = fliplr(xvals);
fill([xvals slavx], [fv(xvals) gv(slavx)], [.8 0 1])
```

Now it works properly. Sample output from this M-file is shown in Figure 12.4. Try
it out on the other examples we have discussed, or on others of your choice.

Solutions to the Practice Sets

Solutions to Practice Set A: Algebra and Arithmetic

1.

(a)

```
1111 - 345
ans =

   766
```

(b)

```
format long; [exp(14), 382801*pi]
ans =

   1.0e+06 *

   1.202604284164777   1.202604809386827
```

The second number is bigger.

(c)

```
[2709/1024; 10583/4000; 2024/765]
ans =

   2.645507812500000
   2.645750000000000
   2.645751633986928
```

```
sqrt(7)
ans =

   2.645751311064591
```

The third, that is $2024/765$, is the best approximation.

2.

(a)

```
cosh(0.1)

ans =

   1.005004168055804
```

(b)

```
log(2)

ans =

   0.693147180559945
```

(c)

```
atan(1/2)

ans =

   0.463647609000806
```

```
format short
```

3.

```
[x, y, z] = solve('3*x + 4*y + 5*z = 2', ...
    '2*x - 3*y + 7*z = -1', 'x - 6*y + z = 3', ...
    'x', 'y', 'z')

x =

241/92


y =

-21/92


z =

-91/92
```

Now we'll check the answer.

```
A = [3, 4, 5; 2, -3, 7; 1, -6, 1]; A*[x; y; z]

ans =

    2
   -1
    3
```

It works!

4.

```
[x, y, z] = solve('3*x - 9*y + 8*z = 2', ...
    '2*x - 3*y + 7*z = -1', 'x - 6*y + z = 3', ...
    'x', 'y', 'z')

Warning: The solutions are parametrized by the symbols:
z1 = C_


x =

- (13*z1)/3 - 5/3


y =

- (5*z1)/9 - 7/9


z =

z1
```

We get a one-parameter family of answers depending on the variable z. In fact, the three planes determined by the three linear equations are not independent, because the first equation is the sum of the second and third. The locus of points that satisfy the three equations is not a point, the intersection of three independent planes, but rather a line, the intersection of two distinct planes. Once again we check:

```
B = [3, -9, 8; 2, -3, 7; 1, -6, 1]; B*[x; y; z]

ans =

    2
   -1
    3
```

5.

```
syms x y; factor(x^4 - y^4)
```

```
ans =
```

```
(x - y)*(x + y)*(x^2 + y^2)
```

6.

(a)

```
simplify(1/(1 + 1/(1 + 1/x)))
```

```
ans =
```

```
(x + 1)/(2*x + 1)
```

(b)

```
simplify(cos(x)^2 - sin(x)^2)
```

```
ans =
```

```
cos(2*x)
```

7.

```
3^301
```

```
ans =
```

```
   4.1067e+143
```

```
pretty(sym('3^301'))
```

```
    4106744371757651279739780821462649478993910868760123094144405702351069915324 9\
    7229781400618467066824164751453321793982128440538198297087323698003
```

8.

(a)

```
solve(67*x + 32 == 0)
```

```
ans =
```

```
-32/67
```

(b)

```
vpa(ans, 15)
```

```
ans =

-0.477611940298507
```

(c)

```
syms p q
pretty(solve(x^3 + p*x + q == 0, x))
```

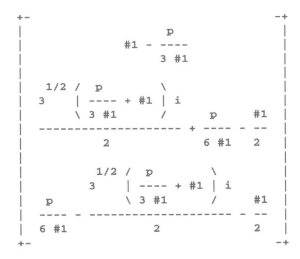

```
                                             where
```

(d)

```
ezplot(exp(x)); hold on; ezplot(8*x - 4)
title('e^x and 8x - 4')
```

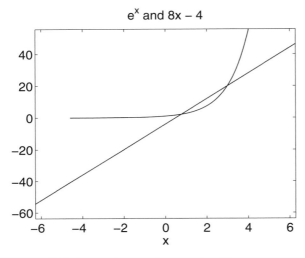

```
fzero(@(x) exp(x) - 8*x + 4, 1)
ans =

    0.7700

fzero(@(x) exp(x) - 8*x + 4, 3)
ans =

    2.9929
```

9.

(a)

```
figure; ezplot(x^3 - x, [-4 4])
```

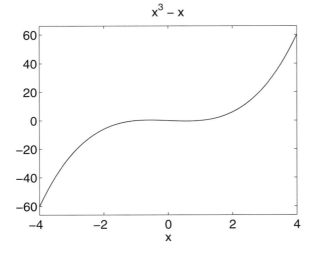

(b)

```
figure; ezplot(sin(1/x^2), [-2 2])
```

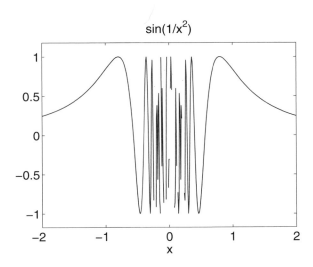

```
figure; X = -2:0.1:2; plot(X, sin(1./X.^2))
```

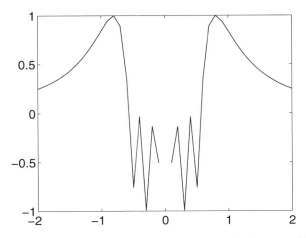

Both pictures are incomplete. Let's see what happens if we refine the mesh.

```
X = -2:0.001:2;
figure; plot(X, sin(1./X.^2))
```

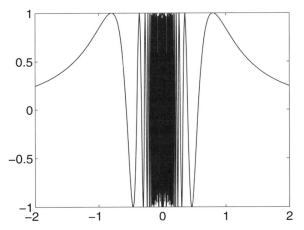

This is better, but because of the wild oscillation near $x = 0$, neither **plot** nor **ezplot** gives a totally accurate graph of the function.

(c)

```
figure; ezplot(tan(x/2), [-pi pi]); axis([-pi, pi, -10, 10])
```

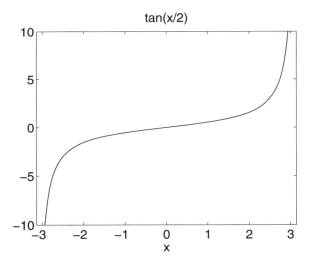

(d)

```
X = -1.5:0.01:1.5;
figure; plot(X, exp(-X.^2), X, X.^4 - X.^2)
```

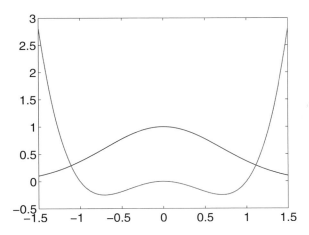

10.

Let's plot 2^x and x^4 and look for points of intersection. We plot them first with **ezplot** just to get a feel for the graph.

```
figure; ezplot(x^4); hold on; ezplot(2^x); hold off
title('x^4 and 2^x')
```

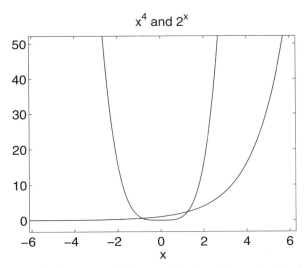

Note the large vertical range. We learn from the plot that there are no points of inter-section between 2 and 6 or between -6 and -2; but there are apparently two points of intersection between -2 and 2. Let's change to **plot** now and focus on the interval between -2 and 2. We'll plot x^4 dashed.

```
X = -2:0.1:2; plot(X, 2.^X);
hold on; plot(X, X.^4, '--'); hold off
```

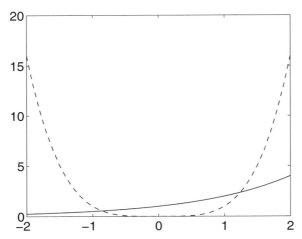

We see that there are points of intersection near -0.9 and 1.2. Are there any other points of intersection? To the left of 0, 2^x is always less than 1, whereas x^4 goes to infinity as $x \to -\infty$. However, both x^4 and 2^x go to infinity as $x \to \infty$, so the graphs might cross again to the right of 6. Let's check:

```
X = 6:0.1:20; plot(X, 2.^X);
hold on; plot(X, X.^4, '--'); hold off
```

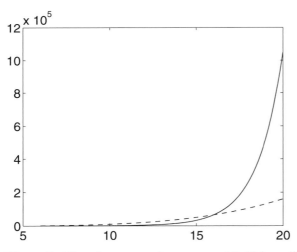

We see that they do cross again, near $x = 16$. Using a little calculus, one can show that the graphs never cross again (by taking logarithms, for example), so we have found all the points of intersection. Now let's use **fzero** to find these points of intersection numerically. This command looks for a solution near a given starting point. To find the three different points of intersection, we will have to use three different starting points. The above graphical analysis suggests appropriate starting points.

```
r1 = fzero(@(x) 2^x - x^4, -0.9)
r2 = fzero(@(x) 2^x - x^4, 1.2)
r3 = fzero(@(x) 2^x - x^4, 16)

r1 =

   -0.8613

r2 =

    1.2396

r3 =

    16
```

Let's check that these "solutions" satisfy the equation.

```
double(subs(2^x - x^4, x, r1))
double(subs(2^x - x^4, x, r2))
subs(2^x - x^4, x, r3)

ans =

    1.5091e-16

ans =

   -5.5107e-16

ans =

0
```

So **r1** and **r2** very nearly satisfy the equation, and **r3** satisfies it exactly. It is easily seen that 16 is a solution. It is also interesting to try **solve** on this equation.

```
syms x real; symroots = solve(2^x - x^4 == 0)

symroots =

   -(4*lambertw(0, log(2)/4))/log(2)
   -(4*lambertw(0, -log(2)/4))/log(2)
  -(4*lambertw(-1, -log(2)/4))/log(2)
```

The function **lambertw** is a "special function" that is built into MATLAB. You can learn more about it by typing **help lambertw**. The command **solve** finds all of the roots, as you can see by looking at the decimal values.

```
double(symroots)
```

```
ans =

   -0.8613
    1.2396
   16.0000
```

Solutions to Practice Set B: Math, Graphics, and Programming

```
clear all
```

1.

(a)

```
[X, Y] = meshgrid(-1:0.1:1, -1:0.1:1);
contour(X, Y, 3*Y + Y.^3 - X.^3, 'k')
```

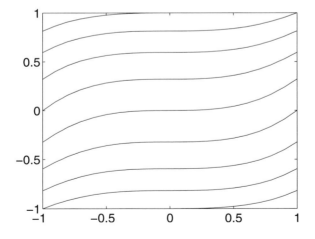

```
[X, Y] = meshgrid(-10:0.1:10, -10:0.1:10);
contour(X, Y, 3*Y + Y.^3 - X.^3, 'k')
```

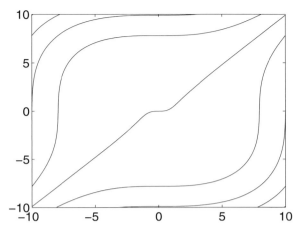

Here is a plot with more level curves. (Note that while **ezcountour** can make the previous plot, it lacks the capability to adjust the number of curves plotted.)

```
[X, Y] = meshgrid(-10:0.1:10, -10:0.1:10);
contour(X, Y, 3*Y + Y.^3 - X.^3, 30, 'k')
```

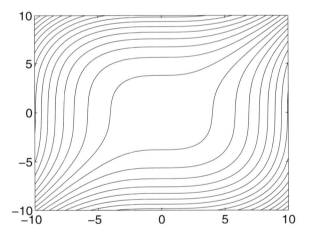

(b)

Now we plot the level curve with value 5.

```
[X, Y] = meshgrid(-5:0.1:5, -5:0.1:5);
contour(X, Y, 3.*Y + Y.^3 - X.^3, [5 5], 'k')
```

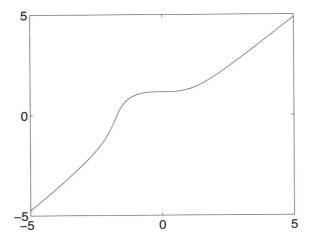

(c)

We note that $f(1, 1) = 0$, so the appropriate commands to plot the level curve of f through the point $(1, 1)$ are:

```
[X, Y] = meshgrid(0:0.1:2, 0:0.1:2);
contour(X, Y, Y.*log(X) + X.*log(Y), [0 0], 'k')
```

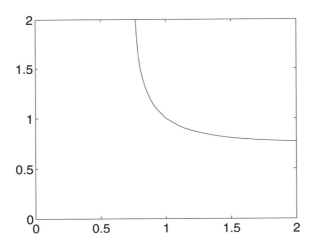

2.

We find the derivatives of the given functions:

```
syms x r
```

(a)

```
diff((2*x - 1)/(x^2 + 1), x)

ans =

2/(x^2 + 1) - (2*x*(2*x - 1))/(x^2 + 1)^2


simplify(ans)

ans =

(2*(- x^2 + x + 1))/(x^2 + 1)^2
```

(b)

```
diff(sin(3*x^2 + 2), x)

ans =

6*x*cos(3*x^2 + 2)
```

(c)

```
diff(x^r, x)

ans =

r*x^(r - 1)
```

3.

We compute the following integrals:

(a)

```
int(cos(x), x, 0, pi/2)

ans =

1
```

(b)

```
int(sin(3*x)*sqrt(1 - cos(3*x)), x)
```

```
ans =
```

```
(2*(1 - cos(3*x))^(3/2))/9
```

```
diff(ans, x)
```

```
ans =
```

```
sin(3*x)*(1 - cos(3*x))^(1/2)
```

(c)

```
int(x^2*sqrt(x + 3), x)
```

```
ans =
```

```
-(2*(x + 3)^(3/2)*(42*x - 5*(x + 3)^2 + 21))/35
```

```
diff(ans, x)
```

```
ans =
```

```
(2*(10*x - 12)*(x + 3)^(3/2))/35 - (3*(x + 3)^(1/2)*(42*x - 5*(x + 3)^2 + 21))/35
```

```
simplify(ans)
```

```
ans =
```

```
x^2*(x + 3)^(1/2)
```

(d)

```
int(exp(-x^2), x, -Inf, Inf)
```

```
ans =
```

```
pi^(1/2)
```

4.

(a)

```
format long; integral(@(x) exp(sin(x)), 0, pi)
```

```
ans =

   6.208758035711110
```

This is an integral MATLAB can't do symbolically:

```
syms x; int(exp(sin(x)), x, 0, pi)
```

```
ans =

int(exp(sin(x)), x == 0..pi)
```

By asking the Symbolic Toolbox to evaluate the integral numerically, we get the same answer as above.

```
double(int(exp(sin(x)), x, 0, pi))
```

```
ans =

   6.208758035711110
```

(b)

```
integral(@(x) sqrt(x.^3 + 1), 0, 1)
```

```
ans =

   1.111447970532575
```

MATLAB can do this integral symbolically, though the exact answer is rather long and cryptic. Here we convert it to a decimal:

```
vpa(int(sqrt(x^3 + 1), x, 0, 1), 20)
```

```
ans =

1.1114479705325754839 + 1.5146129380243426664e-28*i
```

The numerical approximation is accurate to at least 15 decimal places.

(c)

```
integral(@(x) exp(-x.^2), -Inf, Inf)
```

```
ans =
```

```
    1.772453850780313
```

In problem 3(d), we saw that the exact integral is $\sqrt{\pi}$.

```
sqrt(pi)
```

```
ans =
```

```
    1.772453850905516
```

The numerical approximation using **integral** is accurate to 9 decimal places. We can get more accuracy by decreasing the relative error tolerance from its default value of **1e-6**.

```
integral(@(x) exp(-x.^2), -Inf, Inf, 'RelTol', 1e-13)
```

```
ans =
```

```
    1.772453850905516
```

5.

(a)

```
limit((1 + cos(x))/(x + pi), x, -pi)
```

```
ans =
```

```
0
```

(b)

```
limit(x^2*exp(-x), x, Inf)
```

```
ans =
```

```
0
```

(c)

```
limit(1/(x - 1), x, 1, 'left')
```

```
ans =
```

```
-Inf
```

(d)

```
limit(sin(1/x), x, 0, 'right')
```

```
ans =
```

```
NaN
```

The answer **NaN** (for "not a number") means there is no single limiting value. In fact, every real number in the interval between -1 and $+1$ is a "limit point" of $\sin(1/x)$ as x tends to zero. You can see why from the graph of $\sin(1/x)$ on the interval $(0, 1]$.

```
X = 0:0.001:1; plot(X, sin(1./X), 'k')
```

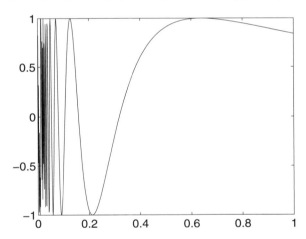

6.

(a)

```
syms k n; symsum(k^2, k, 0, n)
```

```
ans =
```

```
(n*(2*n + 1)*(n + 1))/6
```

(b)

```
syms r; symsum(r^k, k, 0, n)
```

```
ans =
```

```
piecewise([r == 1, n + 1], [r ~= 1, (r*r^n - 1)/(r - 1)])
```

The answer is a little easier to read in "pretty" form.

```
pretty(ans)
```

```
          /                            n                    \
          |                         r r  - 1                 |
piecewise | n + 1 if r == 1,       --------- if r ~= 1      |
          \                         r - 1                    /
```

(c)

```
syms x; symsum(x^k/factorial(k), k, 0, Inf)
```

```
ans =
```

```
exp(x)
```

7.

(a)

```
taylor(exp(x), 'Order', 7)
```

```
ans =
```

```
x^6/720 + x^5/120 + x^4/24 + x^3/6 + x^2/2 + x + 1
```

(b)

```
pretty(taylor(sin(x), 'ExpansionPoint', 2))
```

```
                       2                4
         sin(2) (x - 2)     sin(2) (x - 2)
sin(2) - --------------- + --------------- + cos(2) (x - 2) -
                2                24

            3                5
 cos(2) (x - 2)     cos(2) (x - 2)
 --------------- + ---------------
        6                120
```

(c)

```
taylor(log(x), 'ExpansionPoint', 1, 'Order', 5)

ans =

x - (x - 1)^2/2 + (x - 1)^3/3 - (x - 1)^4/4 - 1
```

8.

(a)

```
syms x y; ezsurf(sin(x)*sin(y), [-3*pi 3*pi -3*pi 3*pi])
```

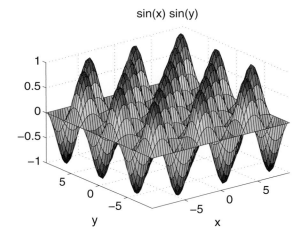

(b)

```
syms x y; ezsurf((x^2 + y^2)*cos(x^2 + y^2), [-1 1 -1 1])
```

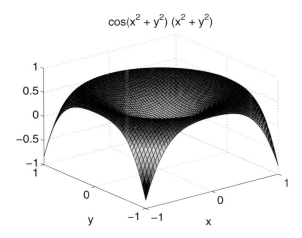

9.

```
T = 0:0.01:1;
for j = 0:16
    fill(4*cos(j*pi/8)+(1/2)*cos(2*pi*T), ...
        4*sin(j*pi/8)+(1/2)*sin(2*pi*T), 'r');
    axis equal; axis([-5 5 -5 5]);
    M(j+1) = getframe;
end
figure(gcf), movie(M)
```

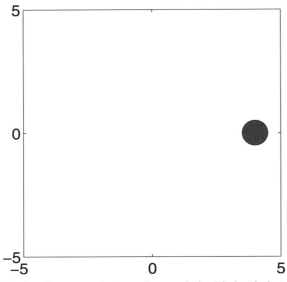

Obviously, we can't show the movie in this book, but you see the last frame above.

10.

(a)

```
A1 = [3 4 5; 2 -3 7; 1 -6 1]; b = [2; -1; 3];
format short; x = A1\b
```

```
x =

    2.6196
   -0.2283
   -0.9891
```

```
A1*x
```

```
ans =

    2.0000
   -1.0000
    3.0000
```

(b)

```
A2 = [3 -9 8; 2 -3 7; 1 -6 1]; b = [2; -1; 3];
x = A2\b
```

```
Warning: Matrix is close to singular or badly scaled. Results may be
inaccurate. RCOND =  4.189521e-18.
```

```
x =

   -6.0000
   -1.3333
    1.0000
```

The matrix **A2** is indeed singular, as we will see in the next problem. The answer given is a valid solution, though as we saw in Problem 4 of Practice Set A, it is not the only solution.

```
A2*x
```

```
ans =

    2
   -1
    3
```

(c)

```
A3 = [1 3 -2 4; -2 3 4 -1; -4 -3 1 2; 2 3 -4 1];
b3 = [1; 1; 1; 1]; x = A3\b3
```

```
x =

   -0.5714
    0.3333
   -0.2857
         0
```

```
A3*x
```

```
ans =

    1.0000
    1.0000
    1.0000
    1.0000
```

(d)

```
syms a b c d u v
A4 = [a b; c d]; A4\[u; v]


ans =

 -(b*v - d*u)/(a*d - b*c)
  (a*v - c*u)/(a*d - b*c)


det(A4)


ans =

a*d - b*c
```

The determinant of the coefficient matrix is the denominator in the solution. So we get a valid solution only if the coefficient matrix is non-singular.

11.

(a)

```
[rank(A1) rank(A2) rank(A3) rank(A4)]


ans =

[ 3, 2, 4, 2]
```

For the fourth matrix, MATLAB implicitly assumes that $ad - bc$ is not 0.

(b)

Only the second matrix is singular.

(c)

```
det(A1)
```

```
ans =

    92
```

```
inv(A1)
```

```
ans =

     0.4239    -0.3696     0.4674
     0.0543    -0.0217    -0.1196
    -0.0978     0.2391    -0.1848
```

```
det(A2)
```

```
ans =

   3.9968e-15
```

```
inv(A2)
```

```
Warning: Matrix is close to singular or badly scaled. Results may be
inaccurate. RCOND =   4.189521e-18.
```

```
ans =

   1.0e+15 *

     9.7578    -9.7578    -9.7578
     1.2510    -1.2510    -1.2510
    -2.2518     2.2518     2.2518
```

MATLAB gives an answer, but notice that the size of the entries is quite large, roughly the inverse of the size of the determinant. The actual determinant of **A2** is 0, and **A2** does not in fact have an inverse.

```
det(sym(A2))
```

```
ans =

0
```

```
inv(sym(A2))
```

```
ans =

FAIL
```

det(A3)

```
ans =

  294.0000
```

inv(A3)

```
ans =

    0.1837   -0.1531   -0.2857   -0.3163
         0    0.1667         0    0.1667
    0.1633    0.0306   -0.1429   -0.3367
    0.2857   -0.0714         0   -0.2143
```

det(A4)

```
ans =

a*d - b*c
```

inv(A4)

```
ans =

[  d/(a*d - b*c), -b/(a*d - b*c)]
[ -c/(a*d - b*c),  a/(a*d - b*c)]
```

12.

(a)

[U1, R1] = eig(A1)

```
U1 =

  -0.9749 + 0.0000i    0.6036 + 0.0000i    0.6036 + 0.0000i
  -0.2003 + 0.0000i    0.0624 + 0.5401i    0.0624 - 0.5401i
   0.0977 + 0.0000i   -0.5522 + 0.1877i   -0.5522 - 0.1877i


R1 =

   3.3206 + 0.0000i    0.0000 + 0.0000i    0.0000 + 0.0000i
   0.0000 + 0.0000i   -1.1603 + 5.1342i    0.0000 + 0.0000i
   0.0000 + 0.0000i    0.0000 + 0.0000i   -1.1603 - 5.1342i
```

A1*U1 - U1*R1

```
ans =

   1.0e-14 *

   0.2665 + 0.0000i    0.1776 - 0.2220i    0.1776 + 0.2220i
  -0.0444 + 0.0000i    0.0888 + 0.0500i    0.0888 - 0.0500i
  -0.0777 + 0.0000i    0.0389 - 0.0888i    0.0389 + 0.0888i
```

This is essentially zero. Notice the **e-14**.

[U2, R2] = eig(A2)

```
U2 =

   0.9669 + 0.0000i    0.7405 + 0.0000i    0.7405 + 0.0000i
   0.1240 + 0.0000i    0.4574 - 0.2848i    0.4574 + 0.2848i
  -0.2231 + 0.0000i    0.2831 + 0.2848i    0.2831 - 0.2848i


R2 =

  -0.0000 + 0.0000i    0.0000 + 0.0000i    0.0000 + 0.0000i
   0.0000 + 0.0000i    0.5000 + 6.5383i    0.0000 + 0.0000i
   0.0000 + 0.0000i    0.0000 + 0.0000i    0.5000 - 6.5383i
```

A2*U2 - U2*R2

```
ans =

   1.0e-14 *

  -0.0685 + 0.0000i    0.0167 - 0.2665i    0.0167 + 0.2665i
  -0.1221 + 0.0000i    0.0444 - 0.1332i    0.0444 + 0.1332i
  -0.0395 + 0.0000i    0.1776 + 0.1776i    0.1776 - 0.1776i
```

Same comment as in (a).

```
[U3, R3] = eig(A3);
```

```
max(abs(A3*U3 - U3*R3))
```

```
ans =

   1.0e-14 *

     0.3770      0.3770      0.2674      0.2674
```

Ditto yet again. (Note that we suppressed the output of the first command, and took the maximum of the absolute values of the entries in each column of **A3*U3 - U3*R3**, in order to keep the data from scrolling off the screen.)

```
[U4, R4] = eig(A4);
pretty(U4)
```

```
pretty(R4)
```

```
simplify(A4*U4 - U4*R4)
```

```
ans =

[ 0, 0]
[ 0, 0]
```

(b)

```
A = [1 0 2; -1 0 4; -1 -1 5];
[U1, R1] = eig(A)

U1 =

   -0.5774    -0.8165     0.7071
   -0.5774    -0.4082    -0.7071
   -0.5774    -0.4082     0.0000


R1 =

    3.0000         0          0
         0    2.0000          0
         0         0     1.0000


B = [5 2 -8; 3 6 -10; 3 3 -7];
[U2, R2] = eig(B)

U2 =

    0.5774     0.8165     0.7071
    0.5774     0.4082    -0.7071
    0.5774     0.4082    -0.0000


R2 =

   -1.0000         0          0
         0    2.0000          0
         0         0     3.0000
```

We observe that the first and second columns of **U1** are negatives of the corresponding columns of **U2**, and the third columns are identical. Finally:

```
A*B - B*A

ans =

     0     0     0
     0     0     0
     0     0     0
```

13.

(a)

If we set X_n to be the column matrix with entries x_n, y_n, and z_n, and M the square matrix with entries $1, 1/4, 0; 0, 1/2, 0; 0, 1/4, 1$, then $X_{n+1} = MX_n$.

(b)

We have $X_n = MX_{n-1} = M^2 X_{n-2} = \ldots = M^n X_0$.

(c)

```
M = [1, 1/4, 0; 0, 1/2, 0; 0, 1/4, 1];
[U,R] = eig(M)


U =

    1.0000         0   -0.4082
         0         0    0.8165
         0    1.0000   -0.4082


R =

    1.0000         0         0
         0    1.0000         0
         0         0    0.5000
```

(d)

M should be URU^{-1}. Let's check:

```
M - U*R*inv(U)


ans =

     0     0     0
     0     0     0
     0     0     0
```

Since R^n is the diagonal matrix with entries $1, 1, 1/2^n$, we know that R_∞ is the diagonal matrix with entries $1, 1, 0$. Therefore, $M_\infty = UR_\infty U^{-1}$. So:

```
Minf = U*diag([1, 1, 0])*inv(U)


Minf =

    1.0000    0.5000         0
         0         0         0
         0    0.5000    1.0000
```

(e)

```
syms x0 y0 z0; X0 = [x0; y0; z0]; Minf*X0

ans =

 x0 + y0/2
         0
 y0/2 + z0
```

Half of the mixed genotype migrates to the dominant genotype and the other half of the mixed genotype migrates to the recessive genotype. These are added to the two original pure types, whose proportions are preserved.

(f)

```
M^5*X0

ans =

 x0 + (31*y0)/64
           y0/32
 (31*y0)/64 + z0
```

```
M^10*X0

ans =

 x0 + (1023*y0)/2048
            y0/1024
 (1023*y0)/2048 + z0
```

(g)

With the suggested alternate model, only the first three columns of the table are relevant, the transition matrix **M** becomes **M = [1 1/2 0; 0 1/2 1; 0 0 0]**. You can compute that the eventual population distribution is **[1 0 0]**, and is independent of the initial population.

14.

The following commands create a JPEG file of the French flag. The **set** command has no effect on the JPEG file; we include it only to turn off the axis ticks in the figure window.

```
A = zeros(200, 300, 3);
A(:, 1:100, 3) = ones(200, 100);
A(:, 101:200, 1) = ones(200, 100);
A(:, 101:200, 2) = ones(200, 100);
A(:, 101:200, 3) = ones(200, 100);
A(:, 201:300, 1) = ones(200, 100);
image(A); axis equal off
imwrite(A, 'tricolore.jpg')
```

We now redefine the color of the leftmost third of the array and create a JPEG file of the Italian flag.

```
A(:, 1:100, 3) = zeros(200, 100);
A(:, 1:100, 2) = ones(200, 100);
image(A); axis equal off
imwrite(A, 'italia.jpg')
```

Of course, you will have to run the commands above yourself to see the flags in color. Then you can run the commands below to see a movie transforming the French flag into the Italian flag.

```
clear M
for k = 0:10
    A(:, 1:100, 3) = (10-k)*ones(200, 100)/10;
    A(:, 1:100, 2) = k*ones(200, 100)/10;
    M(k+1) = im2frame(A);
end
figure(gcf), movie(M)
```

15.

Our solution and its output is below. First, we set **n** to 500 in order to save typing in the following lines and make it easier to change this value later. Then we initialize a zero matrix **A** of the appropriate size and begin a loop that successively defines each row of the matrix. Finally, we extract the maximum value from the list of eigenvalues of **A**.

```
n = 500;
A = zeros(n);
for k = 1:n
    A(k,:) = 1./(k:(k + n - 1));
end
max(eig(A))

ans =

    2.3769
```

There is also a one-line solution using the MATLAB function **hankel**, which creates a matrix constant along anti-diagonals.

```
max(eig(hankel((1:500).^(-1), (500:999).^(-1))))

ans =

    2.3769
```

16.

Again we display below our solution and its output. First, we define a vector t of values between 0 and 2π, in order to be able later to represent circles parametrically as $x = r \cos t, y = r \sin t$. Then, we clear any previous figure that might exist and prepare to create the figure in several steps. Let's say that the red circle will have radius 1; then the first black ring should have inner radius 2 and outer radius 3, and thus the tenth black ring should have inner radius 20 and outer radius 21. We start drawing from the outside in because the idea is to fill the largest circle in black, then fill the next largest circle in white leaving only a ring of black, then fill the next largest circle in black leaving a ring of white, etc. The **if** statement tests true when **r** is odd and false when it is even. We stop the alternation of black and white at a radius of 2 in order to make the last circle red instead of black, then we adjust the axes to make the circles appear round.

```
t = linspace(0, 2*pi, 100);
cla reset; hold on
for r = 21:-1:2
    if mod(r,2)
        fill(r*cos(t), r*sin(t), 'k')
    else
        fill(r*cos(t), r*sin(t), 'w')
    end
end
fill(cos(t), sin(t), 'r')
axis equal; hold off
```

17.

Here are the contents of our solution M-file.

```
type mylcm
```

```
function m = mylcm(varargin)

nums = [varargin{:}];
if ~isnumeric(nums) | any(nums ~= round(real(nums))) | ...
        any(nums <= 0)
    error('Arguments must be positive integers.')
end

for k = 2:length(nums);
    nums(k) = lcm(nums(k), nums(k-1));
end
m = nums(end);
```

Here are some examples. First, cases in which it should work:

```
mylcm(4, 5, 6)

ans =

    60


mylcm([6 7 12 15])

ans =

    420
```

Next, a case in which we expect an error:

```
try
    mylcm(4.5, 6)
catch err
    disp(err.message)
end

Arguments must be positive integers.
```

18.

Here are the contents of our solution M-file.

```
type letcount

function letcount(file)
if isunix
    [stat, str] = unix(['cat ' file]);
else
    [stat, str] = dos(['type ' file]);
end

letters = 'abcdefghijklmnopqrstuvwxyz';
caps = 'ABCDEFGHIJKLMNOPQRSTUVWXYZ';
for n = 1:26
    count(n) = sum(str == letters(n)) + sum(str == caps(n));
end
bar(count)
ylabel 'Number of occurrences'
title(['Letter frequencies in ' file])
set(gca, 'XLim', [0 27], 'XTick', 1:26, 'XTickLabel', ...
    letters')
```

Here is the output when we run this M-file on itself.

letcount('letcount.m')

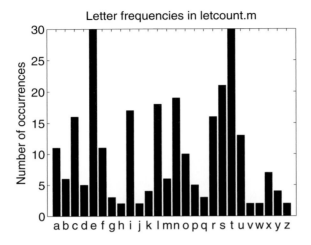

19.

The following M-file performs the desired task:

type editrecent

```
function editrecent

folder = dir('*.m');
if any(size(folder) == 0)
    error('No M-files found.')
end
dates = datenum({folder.date});
[unused, index] = max(dates);
edit(folder(index).name)
```

By assigning the output of **dir** to a variable, we get a structure array that contains among other things the names of all the M-files in the current folder and their modification dates. If the structure array has 0 elements, we print an error message. Otherwise, we form the dates into a cell array of strings, and use **datenum** to convert the strings into numbers. We find the index of the largest number, indicating the most recent date, and finally pass the name of the corresponding file to **edit**.

Solutions to Practice Set C: Developing Your MATLAB Skills

```
clear all
```

1.

(a)

Consider the status of the account on the last day of each month. At the end of the first month, the account has $M + M \times J = M(1 + J)$ dollars. Then at the end of the second month, the account contains $[M(1 + J)](1 + J) = M(1 + J)^2$ dollars. Similarly, at the end of n months, the account will hold $M(1 + J)^n$ dollars. So our formula is

$$T = M(1 + J)^n.$$

(b)

Now we take $M = 0$ and S dollars deposited monthly. At the end of the first month, the account has $S + S \times J = S(1 + J)$ dollars. S dollars are added to that sum the next day, and then at the end of the second month, the account contains $[S(1 + J) + S](1 + J) = S[(1 + J)^2 + (1 + J)]$ dollars. Similarly, at the end of n months, the account will hold

$$S[(1 + J)^n + \cdots + (1 + J)]$$

dollars. We recognize the geometric series – with the leading term missing, so the amount T in the account after n months will equal

$$T = S\left[\frac{(1 + J)^{n+1} - 1}{(1 + J) - 1} - 1\right] = S[((1 + J)^{n+1} - 1)/J - 1].$$

(c)

By combining the two models, it is clear that in an account with an initial balance M and monthly deposits S, the amount of money T after n months is given by

$$T = M(1 + J)^n + S[((1 + J)^{n+1} - 1)/J - 1].$$

(d)

We are asked to solve the equation $(1 + J)^n = 2$ with the values $J = 0.05/12$ and $J = 0.1/12$.

```
syms n
months = double(solve((1 + (0.05)/12)^n == 2));
years = months/12

years =

   13.8918
```

```
months = double(solve((1 + (0.1)/12)^n == 2));
years = months/12

years =

   6.9603
```

If you double the interest rate, you roughly halve the time to achieve the goal.

(e)

```
format bank
1000000/((((1 + 0.08/12)^(35*12 + 1) - 1)/(0.08/12)) - 1)

ans =

     433.06
```

You need to deposit $433.06 every month.

(f)

```
syms n
months = solve(1000000 == ...
    300*(((1 + 0.08/12)^(n+1) - 1)/(0.08/12) - 1));
years = double(months)/12

years =

      39.37
```

You have to work nearly 5 more years.

(g)

First, taking the whole bundle at once, after 20 years the $65,000 left after taxes generates

```
option1 = 65000*(1 + (.05/12))^(12*20)

option1 =

    176321.62
```

The stash grows to about $176,322. The second option yields

```
S = .8*(100000/240);
option2 = S*((1/(.05/12))*(((1 + (.05/12))^241) - 1) - 1)

option2 =

    137582.10
```

You only accumulate $137,582 this way. Taking the lump sum up front is clearly the better strategy.

(h)

```
rates = [.13, .15, -.03, .05, .10, .13, .15, -.03, .05];

clear T
for k = 0:4
    T = 50000;
    for j = 1:5
        T = T*(1 + rates(k + j));
    end
    disp([k + 1,T])
end
```

```
            1.00        72794.74

            2.00        72794.74

            3.00        72794.74

            4.00        72794.74

            5.00        72794.74
```

The results are all the same; you wind up with $72,795 regardless of where you enter in the cycle, because the product $\prod_{1 \leq j \leq 5}(1 + \mathbf{rates}(j))$ is independent of the order in which you place the factors. If you put the $50,000 in a bank account paying 8%, you get

```
50000*(1.08)^5
```

```
ans =
```

```
    73466.40
```

that is \$73,466 – better than the market. The market's volatility hurts you compared to the bank's stability. But of course that assumes you can find a bank that will pay 8%. Now let's see what happens with no stash, but an annual investment instead. The analysis is more subtle here. Set $S = 10,000$. At the end of one year, the account contains $S(1 + r_1)$; then at the end of the second year $(S(1 + r_1) + S)(1 + r_2)$, where we have written r_j for **rates**(j). So at the end of five years, the amount in the account will be the product of S and the number

$$\prod_{j\geq 1}(1 + r_j) + \prod_{j\geq 2}(1 + r_j) + \prod_{j\geq 3}(1 + r_j) + \prod_{j\geq 4}(1 + r_j) + (1 + r_5).$$

If you enter at a different year in the business cycle, the terms get cycled appropriately. So now we can compute

```
format short
clear T TT
for k = 0:4
    T = ones(1, 5);
    for j = 1:5
        TT = 1;
        for l = j:5
            TT = TT*(1 + rates(k + 1));
        end
        T(j) = TT;
    end
    disp([k + 1, sum(T)])
end
```

```
    1.0000     6.1196

    2.0000     6.4000

    3.0000     6.8358

    4.0000     6.1885

    5.0000     6.0192
```

Multiplying each of these by \$10,000 gives the portfolio amounts for the five scenarios. Not surprisingly, all are less than what you obtain by investing the original \$50,000 all at once. But in this model it matters where you enter the business cycle. It's clearly best to start your investment program when a recession is in force and end in a boom. Incidentally, the bank model yields in this case

```
(1/.08)*(((1.08)^6) - 1) - 1

ans =

    6.3359
```

which is better than some investment models and worse than others.

2.

(a)

We can simulate a year in Tony's career by giving him credit for a hit each time a random number between 0 and 1 is ≤ 0.338. The following program does exactly that for n at bats.

type yearbattingaverage

```
function y = yearbattingaverage(n)
% This function file computes Tony's batting average for
% a single year, by simulating $n$ at bats, adding up the
% number of hits, and then dividing by $n$.
hits = 0;
for i = 1:n
    if rand <= 0.338
        hits = hits + 1;
    end
end
y = hits/n;
```

yearbattingaverage(500)

```
ans =

    0.3400
```

(b)

Now let's write a function M-file that simulates a 20-year career. As with the number of at bats in a year, we'll allow for a varying length career.

type career

```
function ave = career(n, k)
% This function file computes the batting average for each
% year in a k-year career, assuming n at bats in each year.
% Then it lists the average for each of the years in the
% career, and finally computes a lifetime average.
Y = zeros(1, k);
for j = 1:k
    Y(j) = sum(rand(n, 1) < 0.338)/n;
end
ave = sum(Y)/k;
disp(['Best avg: ', num2str(max(Y), 4)])
disp(['Worst average: ', num2str(min(Y), 4)])
disp(['Lifetime avg: ', num2str(ave, 4)])
```

career(500, 20)

```
Best avg: 0.376
Worst average: 0.278
Lifetime avg: 0.3388

ans =

    0.3388
```

(c)

If we ran the simulation 100 times and took the average, it would be very close to .338.

3.

We let w, x, y, z denote the number of residences canvassed in the four cities Gotham, Metropolis, Oz, and River City, respectively. Then the linear inequalities specified by the given data are as follows: Non-negative data: $w, x, y, z \geq 0$; Pamphlets: $w + x + y + z \leq 50,000$; Travel cost: $0.5w + 0.5x + y + 2z \leq 40,000$; Time available: $2w + 3x + y + 4z \leq 18,000$; Preferences: $w \leq x$, $x + y \leq z$; Contributions: $w + 0.25x + 0.5y + 3z \geq 10,000$.

The quantity to be maximized is: Voter support: $0.6w + 0.6x + 0.5y + 0.3z$.

(a)

This enables us to set up and solve the linear programming problem in MATLAB as follows:

```
f = [-0.6 -0.6 -0.5 -0.3];
A = [1 1 1 1; 0.5 0.5 1 2; 2 3 1 4; 1 -1 0 0; 0 1 1 -1; ...
    -1 -0.25 -0.5 -3; -1 0 0 0; 0 -1 0 0; 0 0 -1 0; 0 0 0 -1];
b = [50000; 40000; 18000; 0; 0; -10000; 0; 0; 0; 0];
linprog(f, A, b)


ans =

   1.0e+03 *

    1.2683
    1.2683
    1.3171
    2.5854
```

Jane should canvass 1268 residences in each of Gotham and Metropolis, 1317 residences in Oz, and 2585 residences in River City.

(b)

If the allotment for time doubles, then:

```
b = [50000; 40000; 36000; 0; 0; -10000; 0; 0; 0; 0];
linprog(f, A, b)


ans =

   1.0e+03 *

    4.0000
    4.0000
         0
    4.0000
```

Jane should canvass 4000 residences in each of Gotham, Metropolis, and River City, and ignore Oz.

(c)

Finally, if in addition she needs to raise $20,000 in contributions, then

```
b = [50000; 40000; 36000; 0; 0; -20000; 0; 0; 0; 0];
linprog(f, A, b)
```

```
ans =

   1.0e+03 *

    2.5366
    2.5366
    2.6341
    5.1707
```

Jane needs to canvass 2537 residences in each of Gotham and Metropolis, 2634 residences in Oz, and 5171 in River City.

4.

We let w, x, y, z denote the number of hours that Joe spends with the quarterback, the running backs, the receivers, and the linemen, respectively. Then the linear inequalities specified by the given data are as follows: Non-negative data: $w, x, y, z \geq 0$; Time available: $w + x + y + z \leq 50$; Point production: $0.5w + 0.3x + 0.4y + 0.1z \geq 20$; Criticisms: $w + 2x + 3y + 0.5z \leq 75$; Prima Donna status: $x = y$, $w \geq x + y$, $x \geq z$;

The quantity to be maximized is: Personal satisfaction: $0.2w + 0.4x + 0.3y + 0.6z$.

(a)

This enables us to set up and solve the linear programming problem in MATLAB as follows:

```
f = [-0.2 -0.4 -0.3 -0.6];
A = [1 1 1 1; -0.5 -0.3 -0.4 -0.1; 1 2 3 0.5; 0 -1 1 0; ...
     0 1 -1 0; -1 1 1 0; 0 -1 0 1; -1 0 0 0; 0 -1 0 0; ...
     0 0 -1 0; 0 0 0 -1];
b = [50; -20; 75; 0; 0; 0; 0; 0; 0; 0; 0];
linprog(f, A, b)


ans =

   25.9259
    9.2593
    9.2593
    5.5556
```

Joe should spend 9.3 hours each with the running backs and receivers; 5.6 hours with the linemen; and the majority of his time, 25.9 hours, with the quarterback.

(b)

If the team only needs 15 points to win, then:

```
b = [50; -15; 75; 0; 0; 0; 0; 0; 0; 0; 0];
linprog(f, A, b)

ans =

    20.0000
    10.0000
    10.0000
    10.0000
```

Joe can spread his time more evenly, 10 hours each with the running backs, receivers, and linemen; but still the biggest chunk of his time, 20 hours, with the quarterback.

5.

We let f be the quantity to be set to zero. It is a function of the diode voltage, V_D, and the parameters V_0, R, I_0, and V_T. Note that f should be zero for some x between 0 and V_0.

```
f = @(V, V0, R, I0, VT) V - V0 + R*I0*exp(V/VT)

f =

    @(V,V0,R,I0,VT)V-V0+R*I0*exp(V/VT)
```

(a)

```
f1 = @(V) f(V, 1.5, 1000, 10^(-5), .0025)
VD = fzero(f1, [0, 1.5])

f1 =

    @(V)f(V,1.5,1000,10^(-5),.0025)


VD =

    0.0125
```

That's the voltage; the current is therefore

```
I = (1.5 - VD)/1000

I =

    0.0015
```

(b)

```
f2 = @(V) f(V, 1.5, 1000, 10^(-5)/2, .0025);
fzero(f2, [0, 1.5])
```

```
ans =

    0.0142
```

Not surprisingly, the voltage goes up slightly.

(c)

This time, we have to be careful; cutting V_T in half means squaring $\exp(V/V_T)$. So we have to specify an interval big enough to catch the solution, but small enough to prevent the exponential from blowing up too drastically at the right end-point. This will be the case even more dramatically in part (e)

```
f2 = @(V) f(V, 1.5, 1000, 10^(-5), .0025/2);
fzero(f2, [0, 0.5])
```

```
ans =

    0.0063
```

This time the voltage goes down.

(d)

Next we halve both:

```
f2 = @(V) f(V, 1.5, 1000, 10^(-5)/2, .0025/2);
fzero(f2, [0, 0.5])
```

```
ans =

    0.0071
```

The voltage is less than in part (b) but more than in part (c).

(e)

```
X = 10.^(0:-1:-5);
f3 = @(V, x) f(V, 1.5, 1000, 10^(-5)*x, .0025*x);
VD = 0:5; % initialize vector of VD-values
for j = 1:6
    f4 = @(V) f3(V, X(j));
    VD(j) = fzero(f4, [0, X(j)/10]);
end
loglog(10^(-5)*X, VD, 'x-')
xlabel 'I_0'
ylabel 'V_D'
```

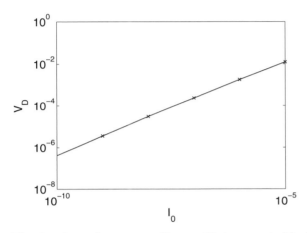

The **loglog** plot appears linear. That suggests V_D is roughly a constant times a power of I_0.

6.

(a)

```
syms x(t) x0 t
dsolve(diff(x) == x - x^2)
```

```
ans =

                  1
                  0
   -1/(exp(C6 - t) - 1)
```

```
sol = dsolve(diff(x) == x - x^2, x(0)== x0)


sol =

-1/(exp(log((x0 - 1)/x0) - t) - 1)
```

When simplified, this includes the constant solutions 0 and 1; indeed:

```
bettersol = simplify(sol)

bettersol =

(x0*exp(t))/(x0*exp(t) - x0 + 1)



subs(bettersol, x0, [0 1])


ans =

[ 0, 1]
```

Let's also check the initial condition.

```
subs(bettersol, t, 0)


ans =

x0
```

b)
```
T = 0:0.1:5;

clf reset, hold on
solcurves = @(t, x0) eval(vectorize(bettersol));
for initval = 0:0.25:2.0
    plot(T, solcurves(T, initval))
end
axis tight
title( ...
'Solutions of dx/dt = x - x^2, with x(0) = 0, 0.25, ..., 2')
xlabel 't'
ylabel 'x'
hold off
```

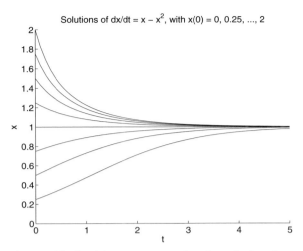

Solutions of $dx/dt = x - x^2$, with $x(0) = 0, 0.25, ..., 2$

The graphical evidence suggests that the solution that starts at zero stays there, and all the others tend toward the constant solution 1.

(c)

```
clf reset, hold on
f = @(t,x) [x(1) - x(1)^2 - 0.5*x(1)*x(2); ...
    x(2) - x(2)^2 - 0.5*x(1)*x(2)];
for a = (0:13)/12, for b = (0:13)/12
        [t, xa] = ode45(f, [0 3], [a,b]);
        plot(xa(:, 1), xa(:, 2))
    end, end
axis([0 13/12 0 13/12])
```

(d)

The endpoints on the curves are the start points. So clearly any curve that starts out inside the first quadrant – that is, one that corresponds to a situation in which both populations are present at the outset – tends toward a unique point, which from the graph appears to be about $(2/3, 2/3)$. In fact, if $x = y = 2/3$, then the right side of both equations in (C.4) vanishes, so the derivatives are zero and the values of $x(t)$ and $y(t)$ remain constant – they don't depend on t. If only one species is present at the outset, that is you start out on one of the axes, then the solution tends toward either $(1, 0)$ or $(0, 1)$, depending on whether x or y is the species present. That is precisely the behavior we saw in part (b).

(e)

```
clf reset, hold on
f = @(t,x) [x(1) - x(1)^2 - 2*x(1)*x(2); ...
    x(2) - x(2)^2 - 2*x(1)*x(2)];
for a = (0:13)/12, for b = (0:13)/12
        [t, xa] = ode45(f, [0 3], [a,b]);
        plot(xa(:, 1), xa(:, 2))
    end, end
axis([0 13/12 0 13/12])
```

This time most of the curves seem to be tending toward one of the points $(1, 0)$ or $(0, 1)$ – in particular, any solution curve that starts on one of the axis (corresponding to no initial population for the other species) does so. It seems that whichever species has a greater population at the outset will eventually take over all the population – the other will die out. But there is a delicate balance in the middle – it appears that, if the two populations are about equal at the outset, then they tend to the unique population distribution at which, if you start there, nothing happens. That value looks like $(1/3, 1/3)$. In fact, that is the value that renders both sides of (C.5) zero – analogous to the role $(2/3, 2/3)$ had in part (d).

(f)

It makes sense to refer to the model (C.4) as "peaceful coexistence," since whatever initial populations you have – provided both are present – you wind up with equal populations eventually. "Doomsday" is an appropriate name for model (C.5), since if you start out with unequal populations, then the smaller group becomes extinct. The lower coefficient 0.5 means relatively small interaction between the species, allowing for coexistence. The larger coefficient 2 means stronger interaction and competition precluding the survival of both.

7.

Here is a Simulink model for redoing the pendulum application:

```
open_system pendulum
```

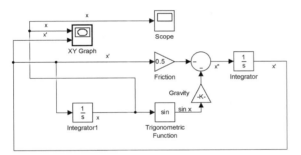

With the initial conditions $x(0) = 0$ and $\dot{x}(0) = 10$, the XY Graph block shows the following phase portrait of x versus $\dot{x}$.

```
pout = sim('pendulum','SaveTime','on','TimeSaveName','t', ...
      'SaveState','on','StateSaveName','x');
x = pout.find('x'); t = pout.find('t');
```

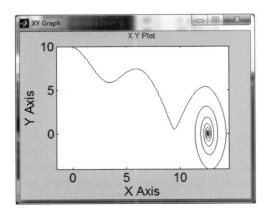

Meanwhile, the Scope block or the following plot gives the following graph of x as a function of t. Note the convergence to 4π (the dotted horizontal line) as $t \to \infty$.

```
plot(t, x(:,1)), hold on
plot([0 30], [4*pi, 4*pi], ':'), hold off
title 'Plot of x as a function of t'
axis([0 30 0 15])
```

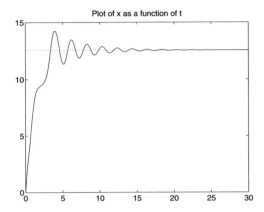

8.

Here is a Simulink model for studying the equation of motion of a baseball.

open_system baseball

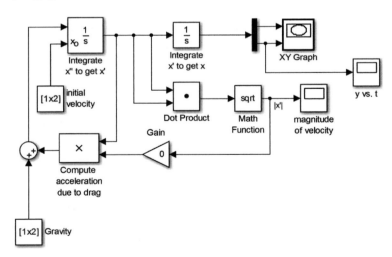

The way this works is as follows. The Integrator block in the upper left integrates the acceleration (a vector quantity) to get the velocity (also a vector). This block requires the initial value of the velocity as an initial condition; we define it in the "initial velocity" Constant block. Output from the first Integrator goes into the second Integrator, which integrates the velocity to get the position (also a vector). The initial condition for the position, `[0, 4]`, is stored in the parameters of this second Integrator. The position vector is fed into a Demux block, which splits off the horizontal and vertical components of the position. These are fed into the XY Graph block, and also the vertical component is fed into a scope block so that we can see the height of the ball as a function of time. The hardest part is the computation of the right-hand side of (C.7). This is computed by adding the two terms on the right with the Sum block near the lower left. The value of `[0, -g]` is stored in the "gravity" Constant block. The second term on the right is computed in the Product block labeled "Compute acceleration due to drag," which multiplies the velocity (a vector) by $-c$ times the speed (a scalar). We compute the speed by taking the dot product of the velocity with itself and then taking the square root; then we multiply by $-c$ in the Gain block in the middle bottom of the model. The Scope block in the lower right plots the ball's speed as a function of time.

(a)

With c set to 0 (no air resistance) and the initial velocity set to $[80, 80]$, the ball follows a familiar parabolic trajectory, as seen in the following picture.

```
bout = sim('baseball','SaveTime','on','TimeSaveName','t', ...
           'SaveState','on','StateSaveName','x');
x = bout.find('x'); t = bout.find('t');
```

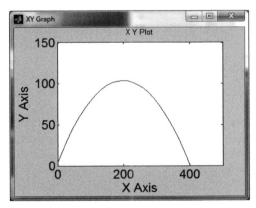

Note that the ball travels about 400 feet before hitting the ground, so the trajectory is just about what is required for a home run in most ballparks. We can read off the flight time and final speed from the other two scopes or plot them as follows:

```
plot(t, x(:,2))
axis([0 10 0 150])
title 'Height versus Time'
```

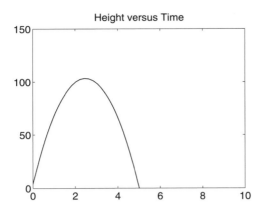

```
plot(t, sqrt(x(:,3).^2 + x(:,4).^2))
axis([0 10 50 300])
title 'Speed versus Time'
```

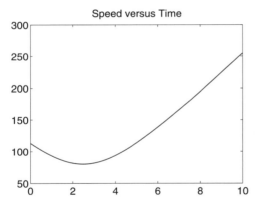

Thus the ball stays in the air about 5.0 seconds and is traveling at about 115 ft/sec when it hits the ground. (Notice that the simulation continues beyond this time; since the model doesn't take into account hitting the ground, the height continues to decrease and the velocity continues to increase as if the ball had been batted off a cliff.)

Now let's see what happens when we factor in air resistance, again with the initial velocity set to $[80, 80]$. First, we take $c = 0.0017$. The trajectory now looks like this:

```
set_param('baseball/Gain', 'Gain', '-0.0017')
bout = sim('baseball','SaveTime','on','TimeSaveName','t', ...
        'SaveState','on','StateSaveName','x');
x = bout.find('x'); t = bout.find('t');
```

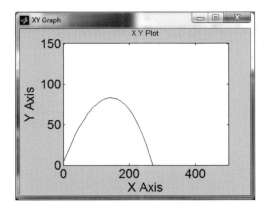

Note the enormous difference air resistance makes; the ball travels only about 270 feet. We can also investigate the flight time and speed with the other two scopes or these plots:

```
plot(t, x(:,2))
axis([0 10 0 150])
title 'Height versus Time'
```

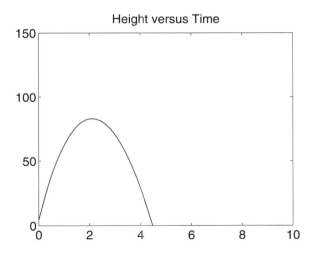

```
plot(t, sqrt(x(:,3).^2 + x(:,4).^2))
axis([0 10 50 300])
title 'Speed versus Time'
```

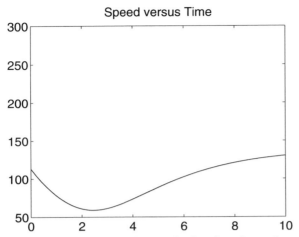

So the ball is about 80 feet up in the air at its peak, and hits the ground in about 4.5 seconds.

(b)

Let's now redo exactly the same calculation with $c = 0.0014$ (corresponding to playing in Denver). The ball's trajectory is now

```
set_param('baseball/Gain', 'Gain', '-0.0014')
bout = sim('baseball','SaveTime','on','TimeSaveName','t', ...
        'SaveState','on','StateSaveName','x');
x = bout.find('x'); t = bout.find('t');
```

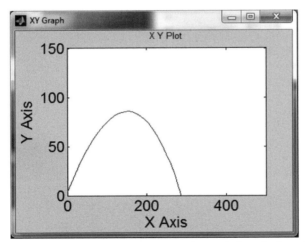

The ball goes about 285 feet, or about 15 feet further than when playing at sea level. This particular ball is probably an easy play, but with some hard-hit balls, those extra 15 feet could mean the difference between an out and a home run. If we look at the height vs. time for the Denver calculation, we see:

```
plot(t, x(:,2))
axis([0 10 0 150])
title 'Height versus Time'
```

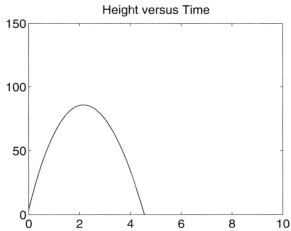

so there is a very small increase in the flight time. Similarly, if we look at the speed vs. time for the Denver calculation, we see:

```
plot(t, sqrt(x(:,3).^2 + x(:,4).^2))
axis([0 10 50 300])
title 'Speed versus Time'
```

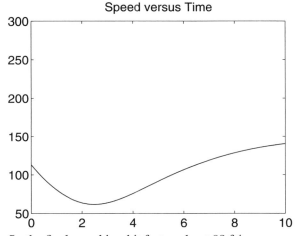

So the final speed is a bit faster, about 83 ft/sec.

(c)

We would expect that batting averages and earned run averages would both be higher in Denver, as indeed is the case according to Major League Baseball statistics.

9.

We construct a Simulink model of the stochastic differential equation.

open_system BS

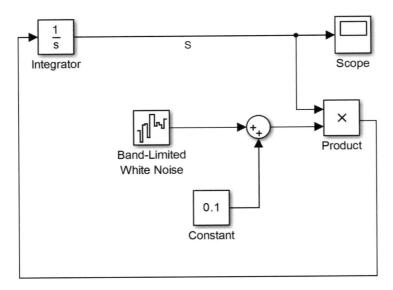

First we run the system with the starting parameters, and plot the result.

```
set_param('BS/Constant', 'Value', '0.1');
set_param('BS/Band-Limited White Noise', 'Cov', '0.2');
BSout = sim('BS','SaveTime','on','TimeSaveName','t', ...
        'SaveState','on','StateSaveName','S');
S = BSout.find('S'); t = BSout.find('t');
plot(t, S)
title 'plot of price vs. time for simulation with noise = 0.2'
```

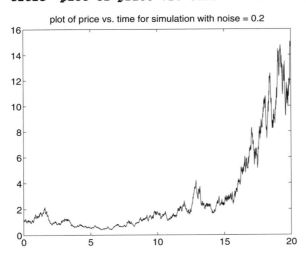

Here is what happens if we increase the constant from 0.1 to 0.15, keeping the noise the same.

```
set_param('BS/Constant', 'Value', '0.15');
BSout = sim('BS','SaveTime','on','TimeSaveName','t', ...
       'SaveState','on','StateSaveName','S');
S = BSout.find('S'); t = BSout.find('t');
plot(t, S)
title 'plot of price vs. time with bigger growth parameter'
```

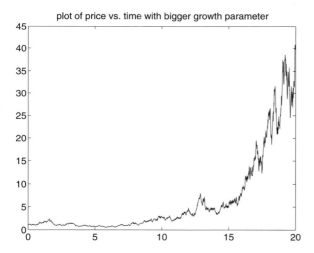

Here is what happens if we increase the noise parameter, going back to the original growth parameter.

```
set_param('BS/Constant', 'Value', '0.1');
set_param('BS/Band-Limited White Noise', 'Cov', '0.3');
BSout = sim('BS','SaveTime','on','TimeSaveName','t', ...
       'SaveState','on','StateSaveName','S');
S = BSout.find('S'); t = BSout.find('t');
plot(t, S)
title 'plot of price vs. time with bigger variance'
```

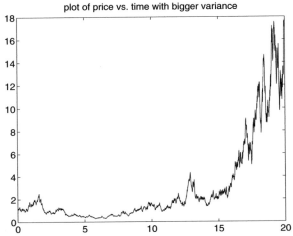

The graphs in this solution look qualitatively similar to some of those in the Financial Data application, though the time scales are different.

10.

Here are the contents of our MuPAD notebook.

C10. Doubling on an Elliptic Curve

This first procedure tests if (x,y) lies on the curve $y^2=x^3+ax+b$. The point at infinity is considered a legitimate point.

```
testEC := (x,y,a,b) -> if (x=infinity and y=infinity)
or x^3+a*x+b-y^2 = 0
then TRUE else FALSE end_if:
```

A few tests:

```
testEC(-1,2,1,6)
```
TRUE
```
testEC(3,6,1,6)
```
TRUE

Doubling of a point P on an elliptic curve E is computed by finding the tangent line L to E through P, finding the other intersection point R of L and E, and then taking -R. (Recall that the negative of a point is its reflection across the x-axis.) The tangent line to E: $y^2=x^3+ax+b$ has slope m given by $2my = 3x^2 + a$, as seen by implicit differentiation. Then if x1 is the x-coordinate of R, it turns out that $x1 = m^2 - 2x$, as we can see from the following:

```
m:=(3*x^2+a)/(2*y):  x1:=m^2-2*x:  y1:=y+m*(x1-x):
simplify(y1^2-x1^3-a*x1)
- x^3 - a*x + y^2
```

And this is just b, since $y^2=x^3+ax+b$. Thus (x1, y1) lies on E!

The next procedure computes elliptic curve doubling.

```
doubleEC := proc(x,y,a,b)
local m, x1;
begin
if not testEC(x,y,a,b)
    then print("point not on curve"); return() end_if;
if (x=infinity and y=infinity)
    then return([x,y]) end_if;
if (y=0)
    then return([infinity,infinity]) end_if;
m := (3*x^2+a)/(2*y):  x1 := m^2 - 2*x:
return([x1, m*(x-x1)-y])
end_proc:
```

Now we do doubling on the curve $y^2=x^3+x+6$.

```
P := [-1, 2];
for j from 1 to 6
    P := doubleEC(P[1], P[2], 1, 6):  print(P);
end_for
```

$[-1, 2]$

$[3, -6]$

$\left[-\frac{5}{9}, -\frac{62}{27}\right]$

$\left[\frac{11131}{8649}, \frac{2468546}{804357}\right]$

$\left[-\frac{21454304639132357}{13176144673437321}, -\frac{35395607922890777 2869506}{1512455244150127732744869}\right]$

[1523679684164295931488248879177869909143078302958941851103 78523835/
412693011789593855634899529580303176730386276110767295488826889, 59\
476052254714258555929797599655592402691778365949143414055157 9911285\
632880365800760845359157997081 8/838379542453350177621652860178510\
10289648800983739452404313214570207052942299375773157591133 8213]

[336858832309982290456352376581016766245844422085887325851180567290\
627403518696600854887391365768576314582806521353107969888691 74225\
402502600974469696224177213304014285559300745915659400376035 6772\
288621343903828470654931561787863332425017374923667077209524 4731/36\
496514666927657010081018453881663602734069706117675727225139 72337\
010896884714417048376930386990794063473691568783420626270315 46093\
588189825493364456437921656808533539988383993395782561087619 23038\
017083790429192079380881623463478215018432496582531012778880 9, 195\
523540650472044672502632853464129396107976465868771473709734 36776\
575376102474276452359262797486372121347035286726670257721397 38722\
790456210523484966676143151791032213727538161343956566362639 30901\
259722469163697716958724201214643554546356880916937667567884 61413\
416548163610846486119529650004605868100861488341535600631691 85175\
635057928031200851734197881886459464167281641433558321459061 74/22\
048400469230671692089979377092528473893421702229418403818602 19220\
750647263296112268421315186175600009267009235124520072985771 44481\
404946592697737543353872983864462111595367183535816475920807 48710\
689934472257831786321315722217497660444330274488162592062281 104839\
373882778376459431164931653192637876502132161923990151088565 54591\
220124385989127572433549215446262775606874762870594058635877]

Note that the signs of the y-coordinates alternate and that the numerators and denominators "explode". Nevertheless, all the points have rational coordinates.

11.

As suggested in the hint, we start by defining an array **z** of complex numbers whose real parts are chosen from the vector **xvals** and whose imaginary parts are chosen from the vector **yvals**.

```
xvals = linspace(-2, 2, 400);
yvals = linspace(-1.5, 1.5, 300);
[X, Y] = meshgrid(xvals, yvals);
Z = X + i*Y;
```

Next we then apply the given function 100 times to each number in the array. By that time, most of the sequences that are going to diverge to infinity have already become larger than the largest floating-point number that MATLAB can store.

```
for k = 1:100
    Z = Z.^2 - 0.75;
end
```

We now use **isfinite** to assign a value of 1 to the values in the array that are still finite floating-point numbers and 0 to those that have diverged. We then use **imagesc** to color the points with ones differently from those with zeros, and select a gray colormap to make the colors white and black. Finally, we use **set** to reverse the vertical axis; without this command, the lower numbers would appear at the top of the graph. This is the default for **image** and **imagesc**, but is not the way mathematical graphs are normally drawn.

```
clf reset
set(gcf, 'Color', 'White')
imagesc(xvals, yvals, isfinite(Z))
colormap(gray)
set(gca, 'YDir', 'normal')
axis equal tight
```

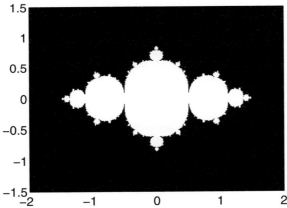

The Julia set consists (approximately) of the boundary between the black and white regions. This boundary is an example of a "fractal," and you can draw a variety of other fractals by changing the expression $z^2 - 0.75$. For example, try changing 0.75 to $0.75 + 0.1i$ and see what a difference it makes!

Glossary

We present here the most commonly used MATLAB objects in six categories: operators, built-in constants, built-in functions, commands, graphics commands, and MATLAB programming constructs. After this we list the most important Simulink commands and blocks, and then we provide a brief list of some of the MuPAD commands referenced in this book. Though MATLAB does not distinguish between commands and functions, it is convenient to think of a MATLAB function as we normally think of mathematical functions. A MATLAB function is something that can be evaluated or plotted; a command is something that manipulates data or expressions or that initiates a process.

We list each operator, function, and command together with a short description of its effect, followed by one or more examples. Many MATLAB commands can appear in a number of different forms, because you can apply them to different kinds of objects. In our examples, we have illustrated the most commonly used forms of the commands. Many commands also have numerous optional arguments; in this Glossary, we have included only some very common options. You can find a full description of all forms of a command, and get a more complete accounting of all the optional arguments available for it, by reading the help text – which you can access by typing either **help <commandname>** or **doc <commandname>**.

This Glossary is not comprehensive. We have selected those MATLAB commands that play a prominent role in this book, or that we believe will be most helpful to the user. You can find a more comprehensive list in the Help Browser.

☞   *See* Online Help *in Chapter 2 for a detailed description of the Help Browser.*

MATLAB Operators

\   Left matrix division. **X = A\B** is the solution of the equation **A*X = B**. Type **help slash** for more information.
      A = [1 0; 2 1]; B = [3; 5];
      A\B

/   Ordinary scalar division, or right matrix division. For matrices, **A/B** is essentially equivalent to **A*inv(B)**. Type **help slash** for more information.

*   Scalar or matrix multiplication. See the online help for **mtimes**.

.   Not a true MATLAB operator. Used in conjunction with arithmetic operators to force element-by-element operations on arrays. Also used to access fields of a structure array.
      a = [1 2 3]; b = [4 -6 8];
      a.*b
      syms x y; solve(x + y - 2, x - y); ans.x

.*   Element-by-element multiplication of arrays. See the previous entry and the online help for **times**.

^   Scalar or matrix powers. See the online help for **mpower**.

.^   Element-by-element powers. See the online help for **power**.

:   Range operator, used for defining vectors and matrices. Type **help colon** for more information.

,   Separates elements of a row of a matrix, or arguments to a command. Can also be used to separate commands on a command line.

;   Suppresses output of a MATLAB command, and can be used to separate commands on a command line. Also used to separate the rows of a matrix or column vector.
```
X = 0:0.1:30;
[1; 2; 3]
```

'   Complex conjugate transpose of a matrix. See **ctranspose**. Also delimits the beginning and end of a string.

.'   Transpose of a matrix. See **transpose**.

...   Line-continuation operator. Cannot be used inside quoted strings. Type **help punct** for more information.
```
1 + 3 + 5 + 7 + 9 + 11 ...
        + 13 + 15 + 17
['This is a way to create very long strings that ', ...
'span more than one line.  Note the square brackets.']
```

!   Runs a command from the operating system. Type **help punct** for more information.

%   Comment. MATLAB will ignore the rest of the same line.

%%   Starts a new cell (if it occurs at the beginning of a line in a script M-file).

@   Creates a function handle or anonymous function.
```
fminbnd(@cos, 0, 2*pi)
f = @(x) x.^2 - x
```

Built-in Constants

eps   Roughly the size of the computer's floating-point round-off error; namely, 2^{-52}, which is approximately 2×10^{-16}.

exp(1) $e = 2.71828\ldots$. Note that **e** has no special meaning.

i $i = \sqrt{-1}$. This assignment can be overridden, e.g., if you want to use **i** as an index in a **for** loop. In that case, **j** can be used for the imaginary unit.

Inf ∞. Also **inf**.

NaN   Not a number. Used for indeterminate expressions like $0/0$.

pi $\pi = 3.14159\ldots$.

Built-in Functions

abs $|x|$.

acos $\arccos x$.

asin $\arcsin x$.

atan $\arctan x$. Use **atan2** instead if you want the angular coordinate θ of the point (x, y).

bessel Bessel functions; **besselj(n, x)** and **bessely(n, x)** are linearly independent solutions of Bessel's equation of order n.

conj Gives the complex conjugate of a complex number.
```
conj(1 - 5*i)
```

cos $\cos x$.

cosh $\cosh x$.

cot $\cot x$.

erf The error function $\mathrm{erf}(x) = (2/\sqrt{\pi}) \int_0^x e^{-t^2} dt$.

exp e^x.

expm Matrix exponential.

gamma The gamma function $\Gamma(x) = \int_0^\infty e^{-t} t^{x-1} dt$ (when Re $x > 0$). The property $\Gamma(k + 1) = k!$, for non-negative integers k, is sometimes useful.

imag $\mathrm{imag}(z)$, the imaginary part of a complex number.

log The natural logarithm $\ln x = \log_e x$.

real $\mathrm{real}(z)$, the real part of a complex number.

sec $\sec x$.

sech $\mathrm{sech}\, x$.

sign Returns -1, 0, or 1, depending on whether the argument is negative, zero, or positive.

sin $\sin x$.

sinh $\sinh x$.

sqrt $\sqrt{x}$.

tan $\tan x$.

tanh $\tanh x$.

zeta The Riemann Zeta function $\zeta(z) = \sum_{n=1}^\infty \frac{1}{n^z}$.

MATLAB Commands

addpath Adds the specified directory to MATLAB's file-search path.
```
addpath C:\my_mfiles
```

ans A variable holding the value of the most recent unassigned output.

cd Makes the specified directory the current (working) directory.
```
cd C:\mydocs\mfiles
```

char Converts a symbolic expression into a string. Useful for defining inline functions.
```
syms x y; f = inline(char(sin(x)*sin(y)))
```

clear Clears values and definitions for variables and functions. If you specify one or more variables, then only those variables are cleared.
```
clear all
clear f g
```

collect Collects coefficients of powers of the specified symbolic variable in a given symbolic expression.
```
syms x y; collect(x^2 - 2*y*x^2 + 3*x + x*y, x)
```

compose   Composition of functions.
```
syms x y; f = exp(x); g = sin(y); h = compose(f, g)
```

ctranspose   Conjugate transpose of a matrix. Usually invoked with the **'** operator. Equivalent to **transpose** for real matrices.
```
A = [1 3 i]
A'
```

D   Not a true MATLAB command. May be used in **dsolve** to denote differentiation. See **diff**.
```
dsolve('x*Dy + y = sin(x)', 'x')
```

delete   Deletes a file.
```
delete <filename>
```

det   The determinant of a matrix.
```
det([1 3; 4 5])
```

diag   Gives a square matrix with a prescribed diagonal vector, or picks out the diagonal in a square matrix.
```
V = [2 3 4 5]; diag(V)
X = [2 3; 4 5]; diag(X)
```

diary   Writes a transcript of a MATLAB session to a file.
```
diary <filename>
diary off
```

diff   Symbolic differentiation operator (also difference operator).
```
syms x; diff(x^3)
diff('x*y^2', 'y')
syms y(x); dsolve(x*diff(y) + y == sin(x))
```

dir   Lists the files in the current working directory. Similar to **ls**.

disp   Displays output without first giving its name.
```
x = 5.6; disp(x)
syms x; disp(x^2)
disp('This will print without quotes.')
```

doc   Opens the Help Browser to documentation on a specific command.
```
doc print
```

double   Gives a double-precision value for either a numerical or a symbolic quantity. Applied to a string, **double** returns a vector of ASCII codes for the characters in the string.
```
z = sym('pi'); double(z)
double('think')
```

dsolve   Symbolic ODE solver. By default, the independent variable is t, but a different variable can be specified as the last argument. The symbolic format is preferred, using **diff** and **==**, but string input is permitted.
```
syms y(x); dsolve(diff(y) + y == 0)
dsolve(diff(y, 2) + y == 0))
dsolve('Dy + y^2 = 0', 'y(0) = 1', 'x')
[x, y] = dsolve('Dx = 2x + y', 'Dy = -x')
```

echo   Turns on or off the echoing of commands inside script M-files.

edit   Opens the specified M-file in the Editor/Debugger.
```
edit mymfile
```

eig Computes eigenvalues and eigenvectors of a square matrix.
```
eig([2, 3; 4, 5])
[e, v] = eig([1, 0, 0; 1, 1, 1; 1, 2, 4])
```

end Last entry of a vector. Also a programming command.
```
v(end)
```

eval Evaluates a string as a MATLAB expression. Useful in M-files.
```
eval('cos(x)')
```

evalin Evaluates a string in the specified workspace. When used with **symengine**, can be used to access the MuPAD kernel.
```
evalin(symengine, 'numlib::fibonacci(100)')
```

expand Expands an algebraic expression.
```
syms x y; expand((x - y)^2)
```

eye The identity matrix of the specified size.
```
eye(5)
```

factor Factors a polynomial or integer.
```
syms x y; factor(x^4 - y^4)
```

feval Evaluates a function specified by a function handle. Useful in function M-files.
```
feval(@exp, 1)
```

find Finds the indices of non-zero elements of a vector or matrix.
```
X = [2 0 5]; find(X)
```

fminbnd Finds the smallest (approximate) value of a function over an interval.
```
fminbnd(@(x) x^4 - x^2 + 1, 0, 1)
```

format Specifies the output format for numerical variables.
```
format long
```

fzero Tries to find a zero of the specified function near a given starting point or on a specified interval.
```
fzero(@(x) cos(x) - x, 1)
fzero(@cos, [-pi 0])
```

guide Opens the GUI Layout Editor.
```
guide mygui
```

help Asks for documentation for a MATLAB command. See also **lookfor**.
```
help factor
```

inline Constructs a MATLAB inline function from a string expression.
```
f = inline('x^5 - x'); f(3)
```

int Integration operator for both definite and indefinite integrals.
```
syms x; int(exp(-x), x, 0, Inf)
int('1/(1 + x^2)', 'x')
```

integral Numerical integration command.
```
format long; integral(@(x) sin(exp(x)), 0, 1)
```

inv Inverse of a square matrix.
```
inv([1 2; 3 5])
```

jacobian Computes the Jacobian matrix, or, for a scalar function, the symbolic gradient.
```
syms x y; f = x^2*y^3; jacobian(f)
sys1=x*(1-x-y); sys2=y*(1-x+2y); jacobian([sys1 sys2])
```

length   Returns the number of elements in a vector or string.
   length('abcde')

limit   Finds a two-sided limit, if it exists. Use **'right'** or **'left'** for one-sided limits.
   syms x; limit(sin(x)/x, x, 0)
   syms x; limit(1/x, x, Inf, 'left')

linspace   Generates a vector of linearly spaced points.
   linspace(0, 2*pi, 30)

load   Loads Workspace variables from a disk file.
   load filename

lookfor   Searches for a specified string in the first line of all M-files found in the MATLAB
   path.
   lookfor ode

ls   Lists files in the current working directory. Similar to **dir**.

max   Computes the arithmetic maximum of the entries of a vector.
   X = [3 5 1 -6 23 -56 100]; max(X)

mean   Computes the arithmetic average of the entries of a vector.
   X = [3 5 1 -6 23 -56 100]; mean(X)

median   Computes the arithmetic median of the entries of a vector.
   X = [3 5 1 -6 23 -56 100]; median(X)

min   Computes the arithmetic minimum of the entries of a vector.
   X = [3 5 1 -6 23 -56 100]; min(X)

more   Turns on (or off) page-by-page scrolling of MATLAB output. Use the SPACE BAR
   to advance to the next page, the RETURN key to advance line-by-line, and Q to abort the
   output.
   more on, help print
   more off

mupad   Opens a MuPAD notebook.
   mupad mynotebook.mn

notebook   Opens an M-Book (Windows only).
   notebook problem1.docx
   notebook -setup

num2str   Converts a number to a string. Useful in programming.
   disp(['The value of pi is ', num2str(pi), '.'])

ode45   Numerical ODE solver for first-order equations. See MATLAB's online help for
   ode45 for a list of other MATLAB ODE solvers.
   [t, y] = ode45(@(t, y) t^2 + y, [0 10], 1); plot(t, y)

ones   Creates a matrix of ones.
   ones(3)
   ones(3, 1)

open   Opens a file. The way this is done depends on the filename extension.
   open myfigure.fig

path   Without an argument, displays the search path. With an argument, sets the search path.
   Type **help path** for details.

pathtool   Opens the "Set Path" tool.

pretty   Displays a symbolic expression in a more readable format.
```
syms x y; expr = x/(x - 3)/(x + 2/y)
pretty(expr)
```

prod   Computes the product of the entries of a vector.
```
X = [3 5 1 -6 23 -56 100]; prod(X)
```

publish   Runs an M-file and "publishes" it. The default is to publish to HTML, but you can specify other formats.
```
publish('mymfile', 'latex')
```

pwd   Shows the name of the current (working) directory.

quit   Terminates a MATLAB session.

rand   Random-number generator; creates arrays of random numbers between 0 and 1.

randn   Normal random-number generator; creates arrays of normal random numbers with mean 0 and variance 1.

rank   Gives the rank of a matrix.
```
A = [2 3 5; 4 6 8]; rank(A)
```

roots   Finds the roots of a polynomial whose coefficients are given by the elements of the vector argument.
```
roots([1 2 2])
```

round   Rounds a number to the nearest integer.

save   Saves Workspace variables to a specified file. See also **diary** and **load**.
```
save filename
```

simple   Attempts to simplify an expression using multiple methods.
```
syms x y;
[expression, how] = simple(sin(x)*cos(y) + cos(x)*sin(y))
```

simplify   Attempts to simplify an expression symbolically.
```
syms x; simplify(1/(1 + x)^2 - 1/(1 - x)^2)
```

size   Returns the number of rows and the number of columns in a matrix.
```
A = [1 3 2; 4 1 5]
[r, c] = size(A)
```

solve   Solves an equation or set of equations. If the right-hand side of the equation is omitted, "0" is assumed. Accepts string input, although symbolic is preferred.
```
syms x y; solve(2*x^2 - 3*x + 6)
[x, y] = solve(x + y == 0, x^2 + y^2 == 1)
[x, y] = solve('x + 3*y = 4', '-x - 5*y = 3', 'x', 'y')
```

sound   Plays a vector through the computer speakers.
```
sound(sin((0:0.1:1000)*pi))
```

str2num   Converts a string to a number. Useful in programming.
```
constant = 'a7'
index = str2num(constant(2))
```

subs   Substitutes for parts of an expression. Accepts string or symbolic input.
```
subs('x^3 - 4*x + 1', 'x', 2)
subs('sin(x)^2 + cos(x)', 'sin(x)', 'z')
syms x y z; subs(z, x, 6)
```

sum   Sums a vector, or sums the columns of a matrix.
```
k = 1:10; sum(k)
```

sym   Creates a symbolic variable or number.

```
sym pi
x = sym('x')
constant = sym('1/2')
```

syms   Defines symbolic variables – **syms x** is equivalent to **x = sym('x')**.

```
syms x y z
syms x real; solve(x^2 + x + 1 == 0)
syms x clear; solve(x^2 + x + 1 == 0)
```

symsum   Performs a symbolic summation of a vector, possibly with infinitely many entries.

```
syms x k n; symsum(x^k, k, 0, n)
syms n; symsum(n^(-2), n, 1, Inf)
```

taylor   Gives a Taylor polynomial approximation with a specified number of terms (the default is 6) at a specified point (default 0). Note: the number of terms includes the constant term, so the default is a polynomial of degree 5, not degree 6.

```
syms x; taylor(cos(x), 'Order', 8)
taylor(exp(1/x), x, Inf, 'Order', 10)
```

transpose   Transpose of a matrix (compare **ctranspose**). Converts a column vector into a row vector, and vice versa. Usually invoked with the **.'** operator.

```
A = [1 3 4]
A.'
```

type   Displays the contents of a specified file.

```
type myfile.m
```

vectorize   Vectorizes a symbolic expression or string. Useful in defining functions.

```
expr = x^2 - 1/x; f = @(x) eval(vectorize(expr))
```

vpa   Evaluates an expression to the specified degree of accuracy using variable-precision arithmetic.

```
vpa('1/3', 20)
```

web   Opens a web browser.

```
web('http://www.mathworks.com')
```

which   Displays the pathname of a command with a given name.

```
which ezplot
which ezplot -all
```

whos   Lists current information on all the variables in the Workspace.

zeros   Creates a matrix of zeros.

```
zeros(10)
zeros(3, 1)
```

Graphics Commands

area   Produces a shaded graph of the area between the x-axis and a curve.

```
X = (0:0.01:4)*pi; Y = sin(X); area(X, Y)
```

axes   Creates an empty figure window.

axis   Sets axis scaling and appearance.

```
axis([xmin xmax ymin ymax])
```
– sets ranges for the axes.

```
axis tight
```
– sets the axis limits to the full range of the data.

axis equal – makes the horizontal and vertical scales equal.

axis square – makes the axis box square.

axis off – hides the axes and tick marks.

bar   Draws a bar graph.
```
bar([2, 7, 1.5, 6])
```

cla   Clears axes.

close   Closes the current figure window. **close all** closes all figure windows.

colormap   Sets the colormap features of the current figure; type **help graph3d** to see examples of colormaps.
```
ezmesh sin(x)*cos(y); colormap cool
```

comet   Displays an animated parametric plot.
```
t = (0:0.01:4)*pi; comet(t.*cos(t), t.*sin(t))
```

contour   Plots the level curves of a function of two variables; usually used with **meshgrid**.
```
[X, Y] = meshgrid(-3:0.1:3, -3:0.1:3);
contour(X, Y, X.^2 - Y.^2)
```

contourf   Filled contour plot. Often used with **colormap**.
```
[X,Y] = meshgrid(-2:0.1:2, -2:0.1:2);
contourf(X, Y, X.^2 - Y.^3);
colormap autumn
```

ezcontour   Easy plot command for contour or level curves.
```
ezcontour('x^2 - y^2')
syms x y; ezcontour(x - y^2)
```

ezmesh   Easy plot command for mesh view of surfaces.
```
ezmesh('x^2 + y^2')
syms x y; ezmesh(x*y)
```

ezplot   Easy plot command for symbolic expressions.
```
ezplot('exp(-x^2)', [-5, 5])
syms x; ezplot(sin(x))
```

ezplot3   Easy plot command for 3D parametric curves.
```
ezplot3('cos(t)', 'sin(t)', 't')
syms t; ezplot3(1 - cos(t), t - sin(t), t, [0 4*pi])
```

ezsurf   Easy plot command for standard shaded view of surfaces.
```
ezsurf('(x^2 + y^2)*exp(-(x^2 + y^2))')
syms x y; ezsurf(sin(x*y), [-pi pi -pi pi])
```

figure   Creates a new figure window.

fill   Creates a filled polygon. See also **patch**.
```
fill([0 1 1 0], [0 0 1 1], 'b'); axis equal tight
```

findobj   Finds graphics objects with specified property values.
```
findobj('Type', 'Line')
```

gca   Gets current axes.

gcf   Gets current figure.

get   Gets properties of a figure.
```
get(gcf)
```

getframe   Command to get the frames of a movie or animation.
```
T = (0:0.01:2)*pi;
for j = 1:12
    plot(5*cos(j*pi/6) + cos(T), 5*sin(j*pi/6) + sin(T));
    axis([-6 6 -6 6]);
    M(j) = getframe;
end
movie(M)
```

ginput   Gathers coordinates from a figure using the mouse (press the ENTER or RETURN key to finish).
```
[X, Y] = ginput
```

grid   Puts a grid on a figure.

gtext   Places a text label using the mouse.
```
gtext('Region of instability')
```

hist   Draws a histogram.
```
hist(rand(200, 1))
```

hold   Holds the current graph. Superimpose any new graphics generated by MATLAB on top of the current figure.
```
hold on
hold off
```

image   Displays a matrix as an image.
```
image(ones(50,100))
```

imagesc   Like **image**, but scales the data if necessary.
```
imagesc(randn(50,100))
```

imread   Reads in a graphics file and converts it to a matrix.
```
A = imread('myimage.jpg');
```

imwrite   Converts a matrix into a graphics file.
```
imwrite(A, 'picture.jpg')
```

legend   Creates a legend for a figure.
```
t = 0:0.1:2*pi;
plot(t, cos(t), t, sin(t))
legend('cos(t)', 'sin(t)')
```

loglog   Creates a log-log plot.
```
x = 0.0001:0.1:12; loglog(x, x.^5)
```

mesh   Draws a mesh surface.
```
[X,Y] = meshgrid(-2:.1:2, -2:.1:2);
mesh(X, Y, sin(pi*X).*cos(pi*Y))
```

meshgrid   Creates a vector array that can be used as input to a graphics command, for example, **contour**, **quiver**, or **surf**.
```
[X, Y] = meshgrid(0:0.1:1, 0:0.1:2);
contour(X, Y, X.^2 + Y.^2)
```

movie   Plays back a movie. See the entry for **getframe**.

movieview   Similar to **movie**, but has a playback button.

patch   Creates a filled polygon or colored surface patch. See also **fill**.
```
t = (0:1:5)*2*pi/5; patch(cos(t), sin(t), 'r'); axis equal
```

pie Draws a pie plot of the data in a vector.
```
Z = [34 5 32 6]; pie(Z)
```

plot Plots vectors of data.
```
X = [0:0.1:2];
plot(X, X.^3)
```

plot3 Plots curves in three-dimensional space.
```
t = [0:0.1:30];
plot3(t, t.*cos(t), t.*sin(t))
```

polar Polar-coordinate plot command.
```
theta = (0:0.01:2)*pi; rho = theta; polar(theta, rho)
```

print Sends the contents of the current figure window to the printer or to a file.
```
print
print -deps picture.eps
```

quiver Plots a (numerical) vector field in the plane.
```
[x, y] = meshgrid(-4:0.5:4, -4:0.5:4);
quiver(x, y, x.*(y - 2), y.*x); axis tight
```

semilogy Creates a semi-log plot, with the logarithmic scale along the vertical axis.
```
x = 0:0.1:12; semilogy(x, exp(x))
```

set Sets properties of a figure.
```
set(gcf, 'Color', [0, 0.8, 0.8])
```

subplot Breaks the figure window into a grid of smaller plots.
```
subplot(2, 2, 1), ezplot('x^2')
subplot(2, 2, 2), ezplot('x^3')
subplot(2, 2, 3), ezplot('x^4')
subplot(2, 2, 4), ezplot('x^5')
```

surf Draws a solid surface.
```
[X,Y] = meshgrid(-2:.1:2, -2:.1:2);
surf(X, Y, sin(pi*X).*cos(pi*Y))
```

text Annotates a figure, by placing text at specified coordinates.
```
syms x; ezplot(x^2); text(0, 0, 'low point')
```

title Assigns a title to the current figure window.
```
title 'Nice Picture'
```

xlabel Assigns a label to the horizontal coordinate axis.
```
xlabel('Year')
```

ylabel Assigns a label to the vertical coordinate axis.
```
ylabel('Population')
```

view Specifies a point from which to view a 3D graph.
```
ezsurf('(x^2 + y^2)*exp(-(x^2 + y^2))'); view([0 0 1])
syms x y; ezmesh(x*y); view([1 0 0])
```

zoom Rescales a figure by a specified factor; **zoom** by itself enables use of the mouse for zooming in or out.
```
zoom
zoom(4)
```

MATLAB Programming

any   True if any element of an array is non-zero.
```
if any(imag(x) ~= 0); error('Inputs must be real.'); end
```

all   True if all the elements of an array are non-zero.

break   Breaks out of a **for** or **while** loop.

case   Used to delimit cases after a **switch** statement.

checkcode   Checks an M-file for common syntax errors and possible coding improvements.
```
checkcode mymfile
```

computer   Outputs the type of computer on which MATLAB is running.

dbclear   Clears breakpoints from a file.
```
dbclear all
```

dbcont   Returns to an M-file after stopping at a breakpoint.

dbstep   Executes an M-file line-by-line after stopping at a breakpoint.

dbstop   Inserts a breakpoint in a file.
```
dbstop in <filename> at <linenumber>
```

dbquit   Terminates an M-file after stopping at a breakpoint.

dos   Runs a command from the operating system, saving the result in a variable. Similar to **unix**.

end   Terminates an **if**, **for**, **while**, or **switch** statement.

else   Alternative in a conditional statement. See **if**.

elseif   Nested alternative in a conditional statement. See the online help for **if**.

error   Displays an error message and aborts execution of an M-file.

find   Reports indices of non-zero elements of an array.
```
n = find(isspace(mystring));
if ~isempty(n)
    firstword = mystring(1:n(1)-1);
    restofstring = mystring(n(1)+1:end);
end
```

for   Begins a loop. Must be terminated by **end**.
```
close; axes; hold on
t = -1:0.05:1;
for k = 0:10
    plot(t, t.^k)
end
```

function   Used on the first line of an M-file to make it a function M-file.
```
function y = myfunction(x)
```

if   Allows conditional execution of MATLAB statements. Must be terminated by **end**.
```
if (x >= 0)
    sqrt(x)
else
    error('Invalid input.')
end
```

input   Prompts for user input.
   answer = input('Please enter [x, y] coordinates: ')

isa   Checks whether an object is of a given class (**double, sym,** etc.).
   isa(x, 'sym')

ischar   True if an array is a character string.

isempty   True if an array is empty.

isequal   Checks whether two arrays are equal. Unlike **==**, does not produce an error message if the arrays have different sizes.
   isequal([1 2], [1 2 3])

isfinite   Checks whether elements of an array are finite.
   isfinite([1 0 1]./[1 0 0])

ishold   True if **hold on** is in effect.

isinf   Checks whether elements of an array are infinite.
   isinf([1 0 1]./[1 0 0])

isletter   Checks whether elements of a string are letters of the alphabet.
   str = 'remove my spaces'; str(isletter(str))

isnan   Checks whether elements of an array are "not-a-number" (which results from indeterminate forms like $0/0$).
   isnan([-1 0 1]/0)

isnumeric   True if an object is of a numerical class.

ispc   True if MATLAB is running on a Windows computer.

isreal   True if an array consists only of real numbers.

isspace   Checks whether elements of a string are spaces, tabs, etc.

isunix   True if MATLAB is running on a UNIX computer.

keyboard   Returns control from an M-file to the keyboard. Useful for debugging M-files.

mex   Compiles a MEX program.

nargin   Returns the number of input arguments passed to a function M-file.
   if (nargin < 2); error('Wrong number of arguments'); end

nargout   Returns the number of output arguments requested from a function M-file.

otherwise   Used to delimit an alternative case after a **switch** statement.

pause   Suspends execution of an M-file until the user presses a key.

return   Terminates execution of an M-file early, or returns to an M-file after **keyboard**.
   if abs(err) < tol; return; end

switch   Alternative to **if** that allows branching to more than two cases. Must be terminated by **end**.

```
switch num
case 1
    disp('Yes.')
case 0
    disp('No.')
otherwise
    disp('Maybe.')
end
```

unix   Runs a command from the operating system, saving the result in a variable. Similar to **dos**.

varargin   Used in a function M-file to handle a variable number of inputs.

varargout   Used in a function M-file to allow a variable number of outputs.

warning   Displays a warning message.

```
warning('Taking the square root of negative number.')
```

while   Repeats a block of commands until a condition fails to be met. Must be terminated by **end**.

```
mysum = 0;
x = 1;
while x > eps
    mysum = mysum + x;
    x = x/2;
end
mysum
```

Simulink Commands

find   Gets values of simulation results

```
x = simOut.find('x');
```

find_system   Finds name of desired Simulink system.

```
find_system(gcs, 'Type', 'block')
```

gcs   "Get current system." Returns name of current Simulink model.

get_param   Gets parameters of a Simulink block or system.

open_system   Opens a Simulink model.

set_param   Sets parameters of a Simulink block or system.

```
set_param('model/Constant', 'Value', '5')
```

sim   Runs a Simulink model.

```
sim('model', 'StopTime', '20')
```

simlp   Solves a linear programming problem.

```
simlp([-1 -1], [1 4; 2 3; 4 1], [3; 2; 3])
```

simplot   Plots data in the style of a Simulink Scope window. Similar to **plot** in most respects.

simulink   Opens the Simulink library.

Simulink Blocks

Algebraic Constraint (Math Operations Library) Equation solver like **fzero**.

Band-Limited White Noise (Sources Library) Generates normally distributed random numbers as in "white noise" sources.

Clock (Sources Library) Outputs the simulation time.

Constant (Sources Library) Outputs a specified constant.

Demux (Signal Routing Library) Disassembles a vector signal into its components.

From Workspace (Sources Library) Takes input from the MATLAB Workspace.

Gain (Math Operations Library) Multiplies by a constant or a constant matrix.

Integrator (Continuous Library) Computes the definite integral, using a specified initial condition.

Math Function (Math Operations Library) Computes exponentials, logarithms, etc.

Mux (Signal Routing Library) Assembles scalar signals into a vector signal.

Polynomial (Math Operations Library) Evaluates a polynomial function.

Product (Math Operations Library) Multiplies or divides signals. Can also invert matrix signals.

Ramp (Sources Library) Outputs a function that is initially constant and then increases linearly starting at the specified time.

Scope (Sinks Library) Plots a signal as a function of time.

Sine Wave (Sources Library) Outputs a sine wave. You can adjust the amplitude, frequency, and phase.

Sum (Math Operations Library) Adds or subtracts inputs.

To Workspace (Sinks Library) Outputs a signal to the Workspace.

Trigonometric Function (Math Operations Library) Computes trigonometric or hyperbolic functions.

Unit Delay (Discrete Library) Useful in modeling difference or differential/difference equations.

XY Graph (Sinks Library) Plots one signal against another.

MuPAD Commands

arctan The arctangent function, which differs from the MATLAB version, **atan**. Note that π in MuPAD is represented by **PI**, which differs from the MATLAB version, **pi**.
```
arctan(1), tan(PI)
```

diff Differentiation operator.
```
diff(x^2, x)
diff(sin(x*y), x)
```

fact The factorial function.
```
fact(23)
```

float Converts an expression to floating point.
```
float(PI)
```

for Introduces a loop.
```
n:=1:
for i from 1 to 10 do
   n:= n*i
end_for
```

if Conditional.
```
for i from -2 to 2 do
   if i<0 then print(i, "is negative")
      elif i=0 then print(i, "is zero")
      else print(i, "is positive")
   end_if
end_for
```

Im Imaginary part of a complex number.

  `Im(exp(PI*I/3))`

int   Integration operator.
  `int(sin(x+y), x)`

linalg::eigenvalues   Computes the eigenvalues of a matrix.
  `A := matrix([[1, 4, 2], [0, 6, 2], [2, 5, 3]]):`
  `linalg::eigenvalues(A)`

ln   The natural logarithm.
  `ln(exp(1))`

ode   Sets up a differential equation.
  `eq := ode(x^2*y'(x) + x*y(x) = cos(x)/x, y(x))`

plot   Displays or combines plots.
  `plot(plot::Function2d(sin(x), x = 0..2*PI))`
  `plot(plot::Function3d(sin(x)+cos(y),`
   `x = 0..2*PI, y = 0..2*PI))`
  `plot(plot::Implicit2d(y^2-x^3+x, x=-1.5..3.5, y=-5..5))`
  `plot(plot::VectorField2d([1, sin(x) + cos(y)],`
   `x = 0..6, y = 0..2.5, Mesh = [31, 26]))`

powermod   Computes a^b mod c.
  `powermod(5, 6, 7)`

proc   Defines a procedure.
  `outputpos := proc(list) begin`
   `select(list, x -> (x>0))`
  `end_proc:`
  `outputpos([i $ i = -10..10])`

Re   Real part of a complex number.
  `Re(1+I)`

solve   Solves an algebraic or differential equation.
  `solve(x^2+ x + 1 = 0, x)`
  `solve(x^2 + y^3 = 1, x)`
  `solve(eq)` – see **ode** above.

subs   Substitution; but the syntax differs from the corresponding MATLAB command.
  `f:=x^2 + y; subs(f, x=2)`

Index